The LOST FLEET

The LOST FLEET

The Discovery of a Sunken Armada
from the Golden Age of Piracy

BARRY CLIFFORD

WILLIAM MORROW
An Imprint of HarperCollins*Publishers*

HarperCollins books may be purchased for educational, business, or sales promotional use. For information please write: Special Markets Department, HarperCollins Publishers Inc., 10 East 53rd Street, New York, NY 10022.

FIRST EDITION

Printed on acid-free paper

Designed by William Ruoto

Library of Congress Cataloging-in-Publication Data

Clifford, Barry.
The lost fleet: the discovery of a sunken armada from
the golden age of piracy/ Barry Clifford.
p. cm.
ISBN 0-06-019818-4
1. Shipwrecks—Venezuela—Aves Island.
2. Pirates—Caribbean Area—History—17th century.
3. France—History, Naval—17th century.
4. Clifford, Barry. I. Title.

G530 .C6219 2001
987'.54—dc21   2001028758

02 03 04 05 06 WBC/QW 10 9 8 7 6 5 4 3 2 1

This book is dedicated to my father,
Robert F. Clifford Jr.

Contents

Preface

This is a book about pirate hunting. More than twenty years ago, I began a hunt that resulted in the discovery of the first pirate-ship wreck ever discovered and authenticated: the *Whydah,* a pirate ship captained by the legendary buccaneer Sam Bellamy that sank more than 250 years ago off the coast of Cape Cod. The project team and I have been recovering artifacts from this underwater time capsule ever since, and each has its own story to tell—of pirates and their ships, raids, brutal sea battles, and sunken treasures. With the *Whydah,* we've been allowed a few glimpses into the world of pirates, yet with each new discovery another mystery has been uncovered. In search for more answers we embarked on a second quest, this time to the coast of Venezuela.

In 1678, on the reef of Las Aves, "the Birds," a small pirate army was shipwrecked together with the majority of a French fleet. It was one of the most fatal naval catastrophes of its time—more than one thousand pirate and French sailors were rumored to have been killed. The event launched an era in pirate history called "the golden age of piracy," a massive outbreak, almost a maritime revolt, that for the next fifty years shook the crowns and counting houses of Europe and would have a profound effect on the history of the Americas.

This naval disaster, an unprecedented, world-altering, and yet nearly forgotten calamity, is the subject of this book.

When I began my investigations into the world of piracy, I thought a pirate was "a monstrous enemy of all mankind" who made his prisoners walk the plank—either a hook-brandishing, rum-swilling, backstabbing Robert Newton or an acrobatic, dashing, misunderstood Errol Flynn. I didn't know that a third of all pirates were of African descent or that pirate crews carried on a unique social experiment, creating a seaborne society that was fundamentally democratic, egalitarian, fraternal, and libertarian. In such "hell towns" as Tortuga, Libertalia, Roatan, New Providence, and even Penzance in Old England, men found that they could choose their own leaders and that those leaders were far more able than those thrust upon them by birth or lineage.

I have since learned that every one of my childhood images of pirates was wrong and I've had to go through the process of getting past cultural stereotypes and suppressed history to the lives of real men. This "demythification" is often lonely work. People like to cling to their legends and myths. "History," someone once said, "is written by the winners"; I have found that much of it has also been written by Hollywood. In both cases, "history from below"—the history of the common man—has often been lost, obscured, or deliberately suppressed. Sometimes the pages of history have to be wrenched open in much the same way explorers had to force their way into the new worlds they found.

The best way for me to make my way to the "Old World"—the world of history—is by discovering and examining the material remains of the past to see with my own eyes how men of the past lived and died. You could say that my entire life has been directed toward learning about the history of piracy. Who were these rough men and what effect did they have on colonial society? Were they simply criminals or was their depraved behavior a reaction to political and economic oppression—a harbinger of revolution? At Las Aves, Venezuela, some of those questions were going to be answered.

During the weeks and months after the catastrophe at Las Aves, treasure hunters came for the ships' remains, looking for gold or, more practical, for cannon and rigging. Decades later, adventurers returned to search once again for the fleet's treasures. When we prepared our expedition to Las Aves in 1998, however, the prospect of gold was not the lure. To us, a team of divers and shipwreck salvors, it was the veil of history we were about to lift that was the most important issue at hand: encrusted cannon, dishes half-buried in the sand, weapons, and barrels—objects untouched for 320 years. We came to Las Aves to learn about the spark that ignited the golden age of piracy.

We have worked to shed light on a near forgotten chapter in history, but this book is more than just a chronicle of a disaster and its consequences. It is also about the spirit of exploration. For better or worse I have never been able to resist the lure of a lost shipwreck and I have found that there is no antidote for gold fever. You carry the germ until the day you die. This is another part of my quest: I feel a fascination—and a kinship—with the explorers of the past. Some of our project team's work, past, present, and future, follows in the wake of great explorers. You will meet some of them in these pages: famous ones like William Dampier, who, but for a quirk of fate, would now be considered common cutthroats. Others may be less familiar. By today's standards some are heroes, some are villains, some were clearly madmen. I believe, however, that John Masefield summed them all up best:

> They were of that old breed of rover whose port lay always a little farther on; a little beyond the skyline. Their concern was not to preserve life, "but rather to squander it away"; to fling it, like so much oil, into the fire for the pleasure of it going up in a blaze. If they lived riotously let it be urged in their favour that at least they lived. They lived their vision. They were ready to die for what they believed to be worth doing. We think them terrible. Life itself is terrible. But life was not terrible to them; for they were comrades, and comrades and brothers-in-arms are stronger than life. Those who "live at home at ease" may condemn them. They are free to do so. The old buccaneers were happier than they. The buccaneers had comrades, and the strength to live their own lives. They may laugh at those who, lacking that strength, would condemn them with the hate of impotence.

Plus ultra!

The LOST FLEET

1

The French Fleet

A Squadron of stout Ships . . .
—A NEW VOYAGE ROUND THE WORLD
William Dampier

They came from the east, running before the steady trade winds that blew along Venezuela's north coast and the islands of the Netherlands Antilles. Ponderous and beautiful, graceful in their heavy and slow way, the ships drove along under deep topsails and fully bellied courses. The French West Indies fleet—great engines of war, like Hannibal's elephants, but vastly more powerful.

Indeed, those ships, *Le Terrible* of seventy guns and five hundred men, *Le Bellseodur* of seventy guns and four hundred fifty men, *Le Tormant* of sixty-six guns and four hundred men, and the fifteen other battleships of the fleet were among the most deadly fighting machines on earth.[1] On May 11, 1678, they were on their way to Curaçao, the last Dutch outpost in the West Indies, to drive out the Dutch and conquer that island for France and her king, Louis XIV.

The events of the night of May 11, 1678, are described in official reports and memoirs, but perhaps the best account comes from William Dampier. Dampier was a sometime Royal Navy officer who circumnavigated the globe three times, sailed with the pirates of the Caribbean and the Pacific, and chronicled his adventures in the best-

A French fleet at sea

selling book *A New Voyage Round the World*. What Dampier did not witness himself he heard firsthand from men who were there. In Dampier's words, the fleet of French admiral Jean Comte d'Estrées was "a Squadron of stout Ships, very well mann'd. . . ."[2]

By the time the fleet sailed for the Netherlands Antilles, the Franco-Dutch War had technically been ongoing for six years. In reality, the previous century of European history had been little more than one long, protracted war between the major powers—France, Spain, England, the Netherlands—interrupted now and again by shaky peace.

The last of those wars, known as the War of Devolution, had ended in 1668, just four years before the outbreak of the Franco-Dutch War. The Treaty of Aix-la-Chapelle concluded the conflict between France and Spain, which had led to the creation of the anti-French Triple Alliance, composed of the United Provinces, now known as the Netherlands, England, and Sweden. Four years of peace, and now they were at it again.

The French fleet that descended on Curaçao had been preparing for action in the Caribbean for a month. Their preparations were well known in the region and caused no end of anxiety, since no one knew

A French squadron weighing anchor

for certain where they were bound or on what unhappy island they might bring their force to bear.

By April 26, Governor William Stapleton on the British island of Nevis could actually watch the fleet gathering in the harbor at Basseterre, the chief town on the neighboring island of St. Kitts. The sight did not please him. He later reported to the Lords of Trade and Plantations that he "was forced by the clamors and cries of the people to secure the helpless sex, old men and children."[3]

That Governor Stapleton should have been so uncertain about whom the French intended to attack is hardly surprising. Ten years before, England and France had been enemies. Four years before, they had been allies against the Dutch. Who knew where they stood now?

At daybreak on the 27th of April, the French were under way. Their actions indicated that perhaps Governor Stapleton's fears were well-founded. All day long the fleet tacked, back and forth, trying to make headway against a southerly wind, appearing to close on Nevis.

Fortunately for that island, those unweatherly seventeenth-century men-of-war could make no progress. Though they worked to windward for better than twelve hours, the French fleet simply could not sail the few miles between St. Kitts and Nevis.

To the great relief of the English colonists, the French fleet finally gave up trying. Governor Stapleton reported that "about sunset they bore away. [I am] Apprehensive they have gone to Martinique to wait

for further orders or to take in men to attack some part of this government . . ."[4] The target of the French fleet was still a mystery.

There is only one man who we can say with certainty knew the
destination of the fleet, and that was Admiral Jean Comte d'Estrées.
Fifty-four years old in 1678, d'Estrées had been in military service
since he was twenty.

D'Estrées was born in Soleure, in present-day Switzerland. He was
of impeccable lineage, like any officer destined for high command. He
also had the good fortune to be born during an era of almost constant
warfare, when military men could count on regular employment, and
the opportunity for distinction and promotion was high.

Comte d'Estrées' first interest was not the navy. He entered the
French army in 1644 and fought in Flanders for the next three years,
being promoted to colonel of the elite Navarre Regiment by the age

Admiral Jean Comte d'Estrées

of twenty-three. By the time he was thirty-one, he had achieved the rank of lieutenant general.

For all of his rapid promotion, Comte d'Estrées was not the ideal soldier. His courage was never questioned; it was demonstrated amply on many occasions. He was a proud and arrogant man (hardly an anomaly among the aristocracy of France), unpleasant to those who served under him, difficult with his superiors. He was described as "a brave man, but a bad leader, and a worse subordinate."[5]

Not until 1668, after quarreling with his senior commander in the army and subsequently quitting the service, did d'Estrées join the French navy. He had never sailed as anything but a passenger before, but thanks to his years of military service, his connections, and his noble birth, he was made vice admiral of the West Indies only three years after entering the navy.

D'Estrées might have done better to remain on land. At first, his career as a fighting sailor was marked by one defeat after another at the hands of the Dutch. Although failure in so large an operation as a major fleet action can rarely be blamed on an individual, d'Estrées' lack of experience and uninspiring leadership did not help.

In 1676, when Louis XIV sent him against the Dutch in the West Indies, d'Estrées finally began to enjoy some success. In December 1676, he captured the Dutch island of Cayenne, and the following year he took Tobago on his second try.

In these actions, d'Estrées' ships were employed largely as transports and the real fighting took place on land. This may have helped d'Estrées, since a land battle was an altogether more familiar situation for the former army lieutenant general. The French were also able to bring overwhelming numbers to bear against the Dutch in that region.

By May 1678, there was only one Dutch stronghold left, Curaçao. Conquering that one small island was all that the French needed to drive their enemy completely from the Caribbean and to acquire the wealth of those islands, leaving France the dominant power in the West Indies.

Curaçao was more than just a lonely outpost. It was situated not far from the stretch of South American coast from Trinidad to Costa Rica known as the Spanish Main. The island was the *entrepôt,* the central gathering spot for goods and shipping, of Dutch trade in the West Indies. Spain, which had huge wealth but little manufacturing and not much of a distribution system, needed the port as a means of supplying its colonies in America. By replacing the Dutch on the island, France stood to put itself in a pivotal position in the New World.

D'Estrées had no intention of failing in this mission. He had in his company eighteen massive warships mounting in total more than seven hundred cannons and carrying more than four thousand men.

But d'Estrées wanted still more, and so he secured for his use a second fleet, a flotilla manned by some of the hardest, most fearless, and most vicious men on earth. Sailing in company with the French fleet, this second armada was composed of fifteen or so ships with their combined complement numbering around fourteen hundred.

They were mercenaries, pirates, renegades. They were the buccaneers of Tortuga.

Brethren of the Coast

2

The Buccaneers

Oh England is a pleasant place for them that's rich and high,
But England is a cruel place for such poor folks as I;
And such a port for mariners I ne'er shall see again
As the pleasant Isle of Avès, beside the Spanish main.
—"THE LAST BUCCANEER"
Charles Kingsley

The early part of the seventeenth century marked the beginning of the end of Spanish rule in the Caribbean. The Spanish were still the dominant power, their great treasure fleets still sailed, and the wealth of that region supported the government in Madrid. But the Caribbean was no longer a Spanish lake.

In the first few decades after Columbus's discovery of the New World, Hispaniola was the focus of Spanish colonization and settlement. While there was some gold found there, the island failed in the expectations of many of the conquistadors. So, when Hernando Cortez came upon the fabulous riches of the Aztec Empire, Hispaniola was quickly depopulated; cattle and hogs were left behind to fend for themselves.

One of the first groups that came to stay after the Spanish exodus were *les boucaniers*—the buccaneers. The first *boucaniers* were hunters, predominantly French, who came in the first quarter of the seven-

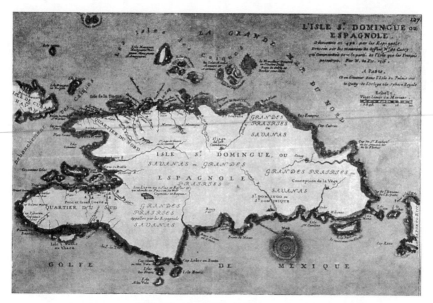

The island of Hispaniola, 1713

teenth century to hunt the feral cattle and pigs the Spanish had abandoned. From the natives they learned the technique of smoking meat on a wooden grill, or *boucan,* to prevent it from spoiling. This smoked meat was not only delicious but very desirable at a time when keeping food edible was a major concern. The *boucaniers* traded their smoked meat for guns, powder, tobacco, liquor, and other essentials with the ships that called at Hispaniola.

In many ways the *boucaniers* of Hispaniola were like the mountain men and trappers who were the first white men into the American west. They were tough, brutal men, not fit for civilized living. Their work was hard and filthy, and they labored in sweltering, mosquito-infested jungles. They were men best left alone.

But Spain, still hoping to maintain absolute control over the rich Caribbean, could not ignore them. Through various means, including slaughtering the animals the buccaneers hunted, the Spanish by the 1630s managed to make Hispaniola untenable for these wild men, who now numbered in the thousands.

Driven from their hunting grounds, the buccaneers settled on the island of Tortuga, just five miles from the northwest coast of Hispaniola.

A buccaneer

Since 1625, Tortuga had been the home of a small French colony, complete with governor and a fortress known as the Dove Côté, though possession of the island shifted back and forth between the French and English colonists and the Spaniards who at various times captured the island, only to be driven off again. Though the island was ostensibly French, as were most of the buccaneers, there was in reality little government there. And since what little French government was there did not much care if the buccaneers were enemies of Spain, it was a fine place for the displaced hunters to make their home.

During their hunting years, the buccaneers had sporadically attacked Spanish shipping, generally when the hunting was not good. Now, deprived of their former livelihood, and with fresh hatred of the Spanish, they began to attack shipping in earnest. In their attempt to eradi-

The fortress of Tortuga

cate the buccaneers, the Spaniards had created a powerful enemy. It was a classic example of the law of unintended consequences.

Early buccaneer successes in attacking the rich homeward-bound Spanish treasure ships encouraged the former hunters to look on piracy as a full-time profession. And they were good at it. Most were excellent shots, grown expert hunting on Hispaniola. They were tough, used to fighting, and had little to lose. Invariably they attacked big ships with big crews, and though outnumbered, the buccaneers were often victorious. From small vessels they graduated to larger and larger ships as they took larger prizes. The early raids on Spanish shipping became a proving ground for these guerrilla warriors.

As the years went on, the wild men who settled on Tortuga became increasingly organized. By the 1640s they had developed a rough pirate democracy, with formalized codes of conduct called the "Custom of the Coast," a form of government that with some variations would be a hallmark of piracy for the next eighty years. They called themselves the Brethren of the Coast.

An attack on a galleon

HELL TOWNS

Piracy is nearly as old as seafaring itself. The word *pirate* comes from a Greek word meaning "sailor." Julius Caesar, as a young man, was captured by pirates. But in the long history of piracy, there have been only a few genuine "hell towns," places that not only catered to pirates but where the population and economy were almost entirely piratical. Port Royal in Jamaica was one such place, as were Nassau on New Providence Island and a number of settlements on the island of Madagascar. Tortuga was the first in the New World.

Tortuga was so bad that around 1650, the French government imported hundreds of prostitutes to the island in an effort to civilize the buccaneers, but this measure had little effect. By 1678, there existed in the Caribbean a genuine outlaw community, a population without a legitimate government. The Brethren of the Coast rejected most aspects of civilized society, submission to authority being first on the list.

The buccaneers tended to operate in small groups, coming together for the purpose of a raiding voyage. They elected their leaders, agreed on the terms of their confederation before setting out, and divided their take evenly.

Despite the antisocial core of the buccaneers' worldview, they were capable of organizing in large numbers. By the time of d'Estrées' attack on Curaçao, a number of leaders had emerged who were able to organize the buccaneers into a sizable force for as long as it took to sack some Spanish town of their choosing.

One of the most successful, and vicious, of the leaders to emerge from the buccaneer community at Tortuga was Francis L'Ollonais. L'Ollonais was a former indentured servant and later a hunter and *boucanier* of Hispaniola who, like many, turned pirate. In 1667 he organized and led an army of seven hundred Tortuga pirates on an attack on Maracaibo in Venezuela.

L'Ollonais also had the distinction of being one of the cruelest and most psychotic of the buccaneers. His fellow pirate Alexandre Exquemelin describes how, frustrated by uncooperative Spanish captives, "L'Ollonais . . . drew out his cutlass, and with it cut open the breast of one of those poor Spaniards, and, pulling out his heart with his sacrilegious hands, began to bite and gnaw it with his teeth."[1]

It was not long after that event that many of L'Ollonais's followers took leave of him. Even for pirates, that was a bit over the top.

The most successful of the buccaneer leaders was Henry Morgan. Morgan was a Welshman who first came to the Caribbean as an English soldier but stayed for a life of piracy. By 1678 he had already coordinated several of these buccaneer armies. In 1668, under Morgan, seven hundred filibusters—that is, freebooters, or pirates—had sailed in a great flotilla and plundered Puerto del Principe in Cuba and Puerto Bello, Panama, in orgies of brutality.

In 1670, Morgan organized a raid on Panama City, collecting together an unprecedented two thousand buccaneers on forty ships. Fighting their way through the rivers and jungles of Panama, they fell on and took the city after defeating a superior number of defenders in a great land battle. Morgan set the standard for the large, organized

Street fighting in a Spanish town

buccaneer raid. He also set the standard for playing the political game, seldom operating without the tacit approval of government authority. Despite his outrageous behavior, Morgan was a favorite in England. He was ultimately knighted and made lieutenant governor of Jamaica.

By the time d'Estrées was ready to attack Curaçao, the precedent of bringing together the buccaneers of Tortuga as a large, amphibious fighting force was well established. The buccaneers, when properly organized, were an effective and devastating weapon. These were the men whom d'Estrées wanted with him.

S.ʳ HEN: MORGAN

Sir Henry Morgan

AN ARMY FOR HIRE

Early in the year 1678, d'Estrées dispatched two frigates to Tortuga
with orders from Louis XIV to the governor, Jacques Nepveu, Sieur
de Pouançay, to raise an ad hoc buccaneer army to join in the attack
on Curaçao.

De Pouançay was able to rally a significant force—between twelve and fourteen hundred buccaneers—no doubt with promises of pay and suggestions of the booty from sacking the Dutch city. The buccaneers brought to the expedition more than a dozen of their own pirate ships, most of which they had captured by staging attacks in smaller vessels. With their fleet they joined the French at Cap François.

None of the buccaneer vessels were close to the size and power of the massive French warships. Still, they were fast and nimble. Their companies, eager and experienced fighting men, were far more effective in battle than the average enlisted soldier or sailor of any nation's regular armed forces. As it happened, the smaller size of the filibusters' ships would be the very thing that would save them.

The fleet of French men-of-war and filibusters sailed from St. Kitts in late April. With the steady trade winds over their larboard quarter, they made their way southwest toward the smattering of islands off the Venezuelan coast. The westernmost three, Aruba, Curaçao, and Bonaire, were the Dutch possessions for which they were bound.

The Venezuelan coast is a treacherous one, and the French navigators were not overly familiar with it, nor were they always in agreement as to where exactly they were. They had no reliable way of determining their longitude, a serious problem in those reef-and-island-strewn waters.

D'Estrées sent a fire ship[2] and three of the smaller filibuster craft several miles ahead of the fleet to scout for navigational hazards. Those ships were more maneuverable than the big French men-of-war, more able to work their way out of any trouble they might get into. Also, they were considerably more expendable.

A fleet of thirty or more ships was not easy to hide, even in the days before radar and airplanes. No doubt the fleet was spotted by passing merchantmen that reported its presence to the governor of Curaçao. While reports differ on this point, the Dutch governor apparently sent out three vessels of no great size to keep an eye on the French fleet but to avoid capture at all costs.

The small Dutch squadron made visual contact with d'Estrées' ships, keeping several miles ahead of them. D'Estrées ordered his lead vessels, the three buccaneer vessels and the fire ship, to go in pursuit of these spies, a perfectly reasonable tactic.

Then, inexplicably, the admiral ordered the rest of the fleet to join in the chase, eighteen big men-of-war going after three small Dutch ships. It was akin to pursuing a mosquito with a sledgehammer, and a

The fire ship

perfect example of the common military blunder of allowing oneself
to become distracted by a sideshow and losing focus on the main
objective.

Who the Dutch captains were or what they were thinking is lost to
history, but we can well imagine their reaction to seeing this massive
fleet coming in pursuit of them. The Dutch mariners, unlike the
French pilots, knew those waters intimately. They knew exactly
where they wanted to go, and they saw a marvelous opportunity.

The chase continued on through the afternoon and into the
evening. The three Dutch ships ran west, with the three buccaneers
and the fire ship in pursuit, and behind those ships, the fleet of Admiral Jean Comte d'Estrées.

Sometime around eight o'clock, with the sun well gone, the Dutch
squadron neared the tiny island of Las Aves. Las Aves was, and is, no
more than a coral outcropping, four miles long with no vegetation to

A French fleet at sea

speak of, and only a few wells dug by pirates who occasionally visited the place. Perhaps the island was visible in the starlight, perhaps not. The French were unaware of the danger into which they were sailing.

What the Dutch knew, and the French did not, was that a great half-moon of submerged reefs, three miles long, ran from the southern tip of the island eastward and then arched away to the north. Three miles of ship-killing rock, perhaps ten feet below the surface, invisible in the dark.

The three Dutch ships passed easily over the reef, as they knew they would. The small buccaneer vessels and the fire ship in pursuit did so as well.

Behind them came the grand French fleet, and foremost in the attack, the bold Comte d'Estrées, eager as ever to get into the fight, plunging recklessly on through the dark.

It was sometime around eight o'clock that the flagship, *Le Terrible*, struck the reef. One can imagine what that moment was like aboard the ship. *Le Terrible* was bowling along under easy sail, a beautiful spring night in the Caribbean, a sure prospect of success for the expe-

A French squadron off of South America

dition. And then in an instant she slammed to a stop, the men thrown off their feet, the heavy bows crushed like eggs, the sick feeling as all aboard realized what had happened. The masts that towered overhead swayed forward and probably broke off at the base, many tons of spars and rigging crashing down to the deck.

In just a minute or two, *Le Terrible* was transformed from one of the most sophisticated, costly, and dangerous fighting machines in the world to mere wreckage.

Comte d'Estrées, not forgetting his duty even at such a moment, ordered cannons fired to warn the rest of the fleet of the danger that lay under their bows. One after another the great guns banged out in the night. To d'Estrées, it was the sound of a warning. To the other captains it was the sound of battle. What none of them knew, or could have known, was that the cannon fire was also the starting signal for the golden age of piracy.

3

Las Aves—Round One

The Las Aves expedition began with a call from Max Kennedy, son of the late Robert Kennedy.

I met Max in Colorado in the late seventies while skiing in Aspen. He must have been twelve or thirteen at the time. His mother, Ethel Kennedy, introduced me as a diver and a shipwreck explorer. That's all Max needed to hear. He followed me around the rest of the day, wanting to learn about everything I had ever done, everything I planned to do in the future, and, most important, could he go with me when I went to do it?

I'd bump into Max on occasion over the years. He never lost his fascination for what lay at the bottom of the sea, or his enthusiasm for shipwreck exploration. In fact, he called me late one night while he was in college and asked if I'd help him plan an expedition to Colombia to hunt for Spanish galleons.

I heard from him again in the fall of 1997. He told me a story about cannon lying on a shallow reef one hundred miles off the coast of Venezuela. He wanted to know if I would help him plan an expedition to find out where they came from.

Although my team and I discovered the wreck of the *Whydah* off Wellfleet, Massachusetts, in 1984, we are still bringing up large quan-

Max Kennedy aboard *Obsession*

tities of artifacts more than fifteen years later. But the excavation season for 1997 was over when Max called. Diving off Cape Cod is
severe in the best of conditions. The water is cold, visibility limited,
and the currents so fierce that excavation pits are filled with sand
almost as soon as they are opened. Tropical reef diving off Venezuela
sounded good.

Max is a strong swimmer and a good diver. If he had been alive in
1492, he would have been the first to volunteer for the Columbus
expedition. And, if that expedition had learned that the world was
indeed flat, Max would have dangled his toes over the edge just for
the fun of it. That's what I like best about Max.

We also share great admiration for the work of historians such as
Stephen Ambrose and the late Samuel Eliot Morison—scholars and
teachers who follow the routes of early explorers in order to test the
accuracy of the primary source record of historically significant events.

After speaking with Max, I consulted with Ken Kinkor, one of the
foremost authorities on the history of piracy, who has been the *Whydah* project historian for the past fifteen years. Ken pursues pirates
through the past the same way the famous manhunter Charles Siringo

pursued the Wild Bunch through the Old West. I occasionally sense that he's looking for something special, but I've never asked him what it is. Tall, infuriatingly deliberate in speech, and with a pipe that emits nearly as much smoke and ash as Mount Saint Helens, Ken might be the perfect model for an especially rumpled classical research professor, were it not for the fact that his chosen subject is far bloodier and far less dignified.

Because Ken is generally familiar with the colonial and maritime history of the West Indies, it didn't take him long to mention some possible wrecks with which the cannon might have been associated. He had certainly heard of the wreck of d'Estrées' fleet, but had not looked into it in detail. With the *Whydah* project at the height of its season, he was too harried to immediately come up with many specifics about the history of the place. As a result, none of us had a full grasp of the enormous potential the reef offered.

Compared with the way we usually do things, we were going in practically blind, but the trip that Max proposed was just reconnaissance in any case. Organized by Max and me, the team consisted largely of Max's friends: Pedro Mezquita, a native Venezuelan who received his law degree from Harvard; film producer Michael Mailer, son of novelist and my Provincetown neighbor Norman Mailer; Michael Karnow, son of historian Stanley Karnow; and Kent Correll, another of Max's friends who is an attorney in New York City.

I hired Chris Macort, a diver who had worked with me for a season on the *Whydah*. Chris was still relatively green, and this would be an important learning experience for him. I put him in charge of packing all the gear for the trip. This is a responsibility that no one really enjoys, but it's essential to the success of any mission. When you are one hundred miles from the nearest hardware store or pharmacy, you had better be sure you have everything you need with you—and multiple items of the most essential equipment. Packing is a fine art, one that is at least as important to me as being a certified diver.

I told Chris, "We're going to explore a reef called 'the Birds,' one hundred miles off the Venezuelan coast." He had a million questions; I told him they would all be answered when we got there, and that I didn't know much more about the place than he did.

Since we weren't sure what we would be up against, we brought everything: standard regulators, Aga masks with regulators, metal detectors, communications gear, bags of spare batteries, and, as always, lots of sharp knives. We also brought tents, sleeping bags, hatchets, and cooking equipment. If we had to camp on the island, we would be prepared.

We also decided to bring along a magnetometer, a long, torpedo-shaped device that is towed behind a boat and detects ferrous anomalies lying on the sea floor. It's clumsy and weighs more than two hundred pounds. This seemed like too much heavy-duty gear for a simple recon. I have found, however, that any equipment you leave behind inevitably proves to be the exact piece of equipment you will need most once you are on-site. In the end, we had nine hundred pounds of gear between the two of us. We were ready for anything—except for the conditions we found.

4

—

A Desolate Place

We arrived in Venezuela on January 8, 1998, and were met by Pedro Mezquita, a big man with a kind face who seemed to laugh, talk, and smile all at the same time. He drove us to a hotel in Caracas, where we stayed before departing for the island.

Max had chartered a boat for us. It was a forty-two-foot Bertram, a power yacht common in marinas in the United States. The term "yacht," with its connotation of luxury, is a little deceptive in this instance, since the stench from the bilge brought vomit to the top of your throat each time you went below.

Las Aves is about a hundred miles off the mainland. It would be a long trip for the little Bertram, Charles Brewer, his wife, their two young children, Chris, and me.

I would guess that Charles is in his early sixties, with the good looks and charm of a successful Rolls-Royce dealer. He is, however, Venezuela's best-known jungle explorer and adventurer.

I am looking a little scruffy at our first meeting aboard the Bertram. "Ah, you look worse than your reputation," he said, introducing me to his wife and kids as "the famous American pirate." Not knowing how I would react, he tried to make it a half-joke with a bad impression of Long John Silver. The contest had begun.

Eric Scharmer, one of our crew, with gear for Las Aves

Educated as a dentist, Charles has impressive credentials. In fact, when he faxed me his résumé, my fax machine ran out of paper after a hundred sheets. Later, he e-mailed me a heavily edited version of twenty-five pages. It started with:

> Curriculum Resume of Charles Brewer-Carius, Explorer and Naturalist Considered the Humboldt of the 20th century by some German publications because of his vast knowledge and experience (in "Inseln in der Zeit p. 275 Uwe George-GEO, 1988), he has developed an overall knowledge of nature, but never the less is very proficient in various fields and has been honored by his fellow scientists naming 26 new species of animals and plants with his name.

It was humbling, especially when I handed him my old business card with a scratched-out phone number. The fact that I "didn't know the difference between a sponge and a gorgonian"—as he would point out later—added to my insecurity.

But it was very easy to like Charles, in spite of his air of superiority, and I remember thinking at the time what great friends we would become if we could get past the testosterone battles. I thought Charles, with his knowledge of the jungle, would be the perfect guide to lead an expedition to the Amazon for a lost Elizabethan shipwreck that I was investigating for the Discovery Channel.

We got under way the next morning. Michael Karnow asked if he could come with us. He knew he was prone to seasickness, and someone had told him that the powerboat would be a smoother ride, although they hadn't factored in the overpowering smell of rotten eggs mixed with fermenting urine in the bilge.

No sooner had we left the protection of the harbor than we were hit by an easterly with the swiftness of a backhand from a cruel stepmother. The little Bertram would climb a wave and then slam down in a great spray of foam. It would be a *long* hundred miles to the "the Island of Birds."

Not having slept since leaving Cape Cod, I went below, stuffed some cotton soaked in mouthwash in my nostrils, jammed myself between two berths, and took a catnap.

When I woke, I found Michael wandering below deck in a state of stupefied agony. I have actually seen dead men with better complexions. He whispered, "I think I'm going to die. . . . No, let me rephrase that: I *want* to die. Would you kindly put me out of my misery?"

I tried to comfort him by reassuring him that *almost* everyone survives seasickness. He tried to laugh, then asked me if I had given his proposal any serious consideration.

The trip took ten hours, and, if given the choice between having a spinal tap or making that crossing again, I'd take the spinal.

Las Aves is a wind-lashed speck of mangroves, bulrushes, and short, coarse grass. It rises from the ocean like a scorpion's tail at the end of a four-mile stretch of deadly reef. As we motored in, it was clear that this island was not a place to camp. Bilge reek notwithstanding, we would stay aboard the little cruiser.

Looking as ghastly as the crew of the *Flying Dutchman,* Max and his group arrived aboard a fifty-foot ketch, even more seasick than our crew. The sailboat had taken the strong easterly on her beam, causing her to wallow with the indignation of an old sow in a muddy ditch.

I arrived aboard the sailboat in the middle of a discussion concerning Charles Brewer. Charles had apparently crossed paths with some of the party before.

"Who invited him along?"

Pedro, ever the lawyer, made a commendable defense on Charles's

Flying out of Las Aves

behalf, and the subject was dropped. But I was left with a lingering suspicion that there was something more to the veteran explorer than I'd seen on his vita.

I took Pedro aside and asked him what the problem was. He hemmed and hawed for a while, but I pressed. Finally, he told me the story.

Some years before, Charles Brewer had led a Kennedy family river expedition in South America that turned into a nightmare and nearly killed Lem Billings, the best friend of the late president John F. Kennedy.

I also heard whispers from some of the boat's crew about Charles, unsettling stories about his time with the Yanomami Indians and other misadventures they had read in the press.

On our first day of the expedition, however, we had problems much greater than personality conflicts. In the Atlantic Ocean, the winds blow in a great clockwise direction, out of the West Indies, up the coast of North America, across the Atlantic to Europe, then down the coast of France, Spain, and Africa. Then, just north of the equator, they return across the Atlantic to the West Indies. These are the trade winds, so called because their predictable direction dictated the trading routes of the old wind-driven ships. When one recalls how d'Estrées tried for an entire day to sail just a few miles into the wind to fetch Nevis, it is obvious why the sailing ships always kept the wind astern.

These steady easterly trade winds, the same winds that drove d'Estrées' fleet relentlessly up on the reefs, were now howling at nearly sixty miles an hour. The wind knocked the boats around, screamed across the cabins, and made even talking in the open difficult. High wind makes everything more complicated and dangerous; it exaggerates all of the potential problems associated with boat work and diving.

Wind generally does not make much difference to a diver under the surface. But the trade winds are relentless. There is nothing between Africa and the coast of Venezuela to slow them down. As the wind blows, it pushes big waves ahead of it, and they smash against the reefs of Las Aves, creating violent waters and strong currents.

The island of Las Aves, the barren hump of land, is to the westward, downwind of the reefs. The reefs stretch out eastward from the bottom corner of the island, completely undetectable to the passing vessel, except for surf. But in these winds, the big seas crash up against the coral, sending spray high into the air, outlining the reefs with a four-mile arc of dangerous breakers.

Stranded atop the reef is a freighter that wrecked fifteen years ago,

Stranded coaster. Note the height of the breakers above the ship.

much like the French fleet. Unlike the wooden men-of-war, however, the freighter did not break up. It is a big ship, perhaps five hundred feet long. The seas that morning were slamming into the side of her decaying hull sixty or seventy feet in the air.

This was not a good situation. In order to get to the wrecks, we had to leave the tranquillity of the lagoon and swim across the reef. The water over the reef was about four feet deep—but there were *waves*. Not the slow, graceful waves you might see in Hawaii, but large, irregular, mutant ones, the size of three-story brownstone houses, which came sporadically and without rhythm, crashing down on the tabletop of poisonous coral.

We took smaller boats to the edge of the reef. After anchoring, my first thought was "There's no way I'm going out there. . . . I'm getting too old for this BS."

And there stood Max, his eyes wide with excitement, as game as my old Labrador to break ice for a crippled goose; some instinct in his blood was calling.

So off we went, one by one, into the stream of rushing water cascading over the reef. The water was only waist deep, but we had to pull ourselves along the bottom, holding on to the coral, to make any progress. It was impossible. The water was just too turbulent. I was

Two crewmen (*foreground*) and (*left to right*) Kent Corell,
Max Kennedy (*on bow*), and Charles Brewer

carrying a metal detector to use once I had reached the other side of the reef. It had earphones to signal if metal was detected. The earphones were being ripped off my head, and, if I turned the detector sideways, the force of the water threatened to tear it out of my hand.

We spent that whole first day banging ourselves against the reef, trying to get over to its seaward side. Disheartened, exhausted, and cut up, we headed back to our boats.

On the morning of the second day, Max and I went out in one of the small boats to take another look. As we were cruising across the lagoon, we saw a small group of men who seemed to be standing on water. As we approached, it became clear that they were conch divers standing on a postage-stamp-size spit of sand. They were happy to see us, as the mother ship that had dropped them off was a week overdue. They were low on food and water, and there was a change of weather in the air.

One of the secrets in finding shipwrecks is local knowledge. These men, if anyone, would know if there were old shipwrecks along the reef. I began to question them in my rudimentary Spanish.

"*Cañones?*" I asked. "Have you seen any cannons?"

"No, no *cañones,* no *cañones,*" they replied. I was disappointed until

Divers investigating wreckage

they went on to explain that, although there were no cannons, they had seen what they described as very large sewer pipes piled up on top of one another.

Sewer pipes. Max and I exchanged glances. While it was possible that they were sewer pipes, these men had probably seen a pile of cannons, all that was left of some ancient shipwreck.

I told them I was very eager to see the sewer pipes.

5

—

"Beasts of Prey"

DEATH ON THE REEF

I may have been eager to get out to the reef, but the sailors and pirates who ended up there three hundred years ago—those who lived through the night—certainly wished they had never seen the place.

Le Terrible piled up on the reefs of Las Aves and instantly became a total loss. Comte d'Estrées tried desperately to prevent the rest of his fleet from meeting the same fate. If he could keep enough of them from running aground, he might still drive the Dutch from Curaçao, even with the loss of the seventy-gun flagship.

Night was fully on them. D'Estrées ordered *Le Terrible's* guns fired and a lantern lit in the main top, the fastest means of signaling the other ships and warning them of the danger. Gunfire proved to be a sadly ambiguous warning. The captains of the great men-of-war astern thought the cannon blasts meant that the admiral was engaged with the enemy.

Their duty was clearly to get up with the flagship and support her.

They ordered more sail set and closed as quickly as they could. As Dampier later reported, the captains "hoisted up their topsails, and crowded all the sail they could make, and ran full sail ashore after him, all within a mile of each other. For his light being in the Main-Top was an unhappy Beacon for them to follow. . . ."[1]

One after another, the great ships ran up on the reef; *Le Tormant* with her four hundred men, *Le Bellseodur* with a complement of four hundred fifty, and *Le Bourbon, Le Prince,* and *Le Hercule,* each with three hundred aboard, all broke their backs on the unseen rocks.

Along with the gunfire, d'Estrées managed to dispatch a boat, which was able to warn the left wing of the fleet in time for those vessels to veer off.[2] But for the ships following in the admiral's wake it was too late.

No doubt some of the men aboard those doomed ships realized the danger before they actually struck the reef, but the unweatherly men-of-war of the late seventeenth century were helpless to sail clear once they had the reefs right under their lee.

Though the process of rigging and dropping the huge anchors was time-consuming, some ships might have managed to do so. But it did not help. The anchors found nothing to grab on the reefs and just dragged along behind. The mariners aboard the French fleet could have done nothing but endure the horror of waiting for the inevitable.

In all, ten French men-of-war were wrecked on Las Aves, the

Loss of a fleet

largest ships in d'Estrées' fleet. Lost with the ships were approximately five hundred men and 490 guns. Even in a major naval engagement, it was rare to see such complete destruction.

Dampier suggests that the buccaneers fared better in this disaster, though the record is unclear. As a renegade mercenary force, the free-booters,[3] as they were called, were of no official concern to the French navy, as long as they did as ordered and aided d'Estrées in his designs. As a result, the navy took no official notice of the number of pirate vessels that were wrecked, and the pirates themselves were not much given to record-keeping.

Some reports that made their way to the colonial governors indicate that as many as eight buccaneer vessels went up on the reef. It is certain that at least three were lost with the men-of-war. The accounts of the aftermath reveal that a number of buccaneers were stranded on the beach along with the survivors from the naval force. Enough of the buccaneer ships escaped, however (many no doubt thanks to their shallow draft), to become a power in their own right.

NIGHTTIME AND BREAKING SEAS

One can only imagine the horror that the shipwrecked mariners suffered through the night of May 11 and 12, with the great men-of-war grinding themselves to bits on the reefs, held there by the trade winds and the steady pounding of the sea.

The air was filled with the sounds of wooden hulls crushing against rock, tons of masts, spars, and rigging collapsing, panicked orders as the officers tried to maintain discipline and salvage what they could of men and ships, the screams of the drowning men, the shudder of waves breaking on the stranded vessels.

Dawn revealed a chaotic scene, with the men-of-war, stove in but still largely intact, stranded on the half-moon reef. The clear blue-green water and the white sand beaches of Las Aves were covered with debris and the bodies of the men who had drowned.

The survivors began the task of getting themselves off the stranded, wrecked ships, which were beginning to break up, and onto the dubious safety of the beaches on Las Aves. The work was difficult. The boats were hampered by large seas that made the generally dangerous task of approaching an unstable wreck even more treacherous.

Lonesome palm

The scene on the beach at Las Aves and on the wrecks still clinging to the reef must have been one of the strangest in all the history of the Caribbean, no small feat in a country where the bizarre and outrageous is a standard part of regional history.

The survivors were divided into two distinct groups: the men of the French navy and the buccaneers. Of the two, the men of the navy had by far the worst time of it.

Most of those killed in the wrecks seem to have been sailors aboard the French men-of-war. This is hardly surprising. Common sailors made up the majority of casualties in any conflict at sea, whether by gunfire or rocks.

All through the day the survivors poured onto the sands of Las Aves. They were frightened and dispirited, and many had been injured in the wreck or cut up by coral in getting ashore. Many more, no doubt, fell victim to the fire coral that grows in those waters.

With the water filled with blood, sharks and barracuda began to school. We saw enough of both, diving at Las Aves, even without the attraction of fresh blood to lure them. The predatory sharks came, silent and unseen, and claimed their own screaming victims.

Once they were on the beach, the sailors' suffering was intensified by the fact that Las Aves was little more than a barren stretch of sand

and shrub, just twelve degrees north of the equator, with no shelter from the elements. The European sailors, unused to such conditions, "died like rotten sheep."[4]

There was at least no shortage of food and drink. As the ships broke up, they began to disgorge their holds full of supplies—casks of beef and pork, water and wine and brandy—all floating free of the wrecks and washing up on the beaches of Las Aves.

Some of the shipwrecked Frenchmen were overcome with despair and reached for the bottle. Forty or so French sailors found themselves aboard a wreck with a good supply of liquor. Rather than trying to save themselves, they chose to get dead drunk. When at last the after part of the ship in which they had settled broke away and floated off, the men could be heard singing merrily as they drifted out to sea, never to be heard from again.

Most of the French naval officers survived but would have to answer for the destruction of the ships under their command. That would not be a pleasant prospect for any of them, least of all Admiral Jean Comte d'Estrées, whose record as a naval commander was mediocre at best. One can imagine d'Estrées looking out over the wreckage and working out the explanation he would give to the Sun King. It must have been the worst moment of his life.

6

A Change of Plans

There were forty craft in Avès, that were both swift and stout,
All furnished well with small arms and cannons round about;
And a thousand men in Avès made laws so fair and free
To choose their valiant captains and obey them loyally.
— *"THE LAST BUCCANEER"*
Charles Kingsley

MAY 12, 1678
LAS AVES

No doubt the officers thrown up on the beach were concerned with more than just their personal reputations and careers. In just a few hours, the bulk of the French naval presence in the Caribbean had been wiped out more completely than any enemy in those waters (least of all the Dutch) could have ever hoped to achieve. Any man of insight could have realized that that single event probably meant the end of any chance for French domination over the West Indies. The Caribbean was, and would continue to be, one of the major sources of European wealth. The loss of the fleet was a major blow to France, the repercussions of which would be felt for nearly a century.

Soon after the disaster, the end of French designs on the region was welcomed with ill-concealed glee among the English colonists in the West Indies. The governor of Barbados, Sir Jonathan Atkins, wrote, "[T]here is little danger now from the French fleet under the command of Count d'Estrées, the greatest part of it having been 'ruined almost to a miracle.' " Atkins went on to add, facetiously, that "d'Estrées is like to give his master a good account of his fleet. I wish them luck at home if we have a war with them."[1]

Comte d'Estrées, true to his character, had lost neither his courage nor his sense of duty. He still held out hopes of driving the Dutch from Curaçao. As a French nobleman, he no doubt cared deeply for French glory and understood the importance of France gaining supremacy in the region. It might also have occurred to him that the blame he would bear for the loss of the fleet could be somewhat mitigated if he could still manage to complete his mission.

With most of his fleet gone, d'Estrées needed the buccaneers more than ever. The pirate ships that survived the reefs, nine or ten in all, now represented half of the admiral's available fleet, and the twelve hundred or so buccaneers a sizable portion of his land forces. Unhappily for d'Estrées, the buccaneers had lost interest in attacking Curaçao.

Concerns of French pride and dominance in the West Indies mattered not a bit to the freebooters. With the big men-of-war wrecked, the odds of sacking Curaçao did not look so attractive, not enough to tempt them into following d'Estrées any longer. As Atkins reported, "D'Estrées lost not his courage, but with the Ships he had left would have attempted [Curaçao], But his Buckneers (which are only Beasts of Prey) seeing there was little to be gott but Blows left him and would not hazard any farther with him."[2]

The buccaneers might have suffered the same hazards as the French sailors in getting ashore, but once on the beach things were much easier for them. These were not sailors fresh from Europe's cool, damp climate, but men who had already been many years in the Caribbean and were acclimated to the heat and the blazing sun. The buccaneers led wild and adventurous lives. Finding themselves shipwrecked on a barren island was not so far removed from the normal course of events.

The circumstances on Las Aves were far better than usual, because a fortune in loot literally drifted in on the tide. The buccaneers discarded their old ragged clothing and donned the finery that had floated free of the cabins, or that they stripped from the bodies of drowned officers. They made tents from the sails torn from shattered yards. They retrieved casks of food and liquor from the surf and rolled them up the beach to their makeshift hell town. One of the survivors of Las Aves

The Buccaneer

later told William Dampier that "if they had gone to Jamaica with 30*l.* a Man in their Pockets, they could not have enjoyed themselves more."[3]

The buccaneers kept to themselves, enjoying the high life at their end of the beach, while at the other end the men of the French navy suffered, wallowed in their despair, and died.

D'Estrées worked hard to rally the freebooters in his camp into proceeding with the attempt on Curaçao, but the men of Tortuga would have none of it. The wreck of the French fleet seemed to have dissolved any ties they felt to the French. The buccaneers felt free to loot whatever they could of their former employers' possessions.

D'Estrées did not remain long on Las Aves. Soon he managed to get the remainder of his force aboard what was left of his fleet. With what they could salvage, they left the scene of the great disaster astern. They did not go to Curaçao, of course, but instead retreated to the French colony of Saint-Domingue. D'Estrées returned to France that summer. By then, the war was all but over.

As thorough and devastating as it was, the loss of the fleet on Las Aves did not completely ruin d'Estrées' reputation or career. Such an accident was much more forgivable at a time when charts were few and mariners had no more than a lead line to warn of dangers below. Louis XIV was more likely to ascribe the disaster to the will of God than a modern head of state might be.

A year later, d'Estrées was back in the Caribbean with yet another fleet under his command. Once again, his presence caused great alarm. In the end, however, he did no more than show the French flag in the region before returning to France. Ironically, he would spend much of his remaining career fighting the Barbary pirates of the North African coast.

Carriage gun that would have been used for the invasion of Curaçao

A GATHERING OF BUCCANEERS

The buccaneers were not so quick to leave Las Aves. They were having a tropical vacation, with all provisions provided free of charge, courtesy of the French navy. They found themselves in a unique position. They were pirates to a man, already assembled and under way, men under arms who no longer had a fight. They had lost only a few ships and a small portion of their company. They still represented a significant fighting force—nine or ten ships and twelve to fifteen hundred men. They had been infected by the dream of plunder, and that dream only needed direction.

The more democratic a society, the more opportunity for natural leaders to emerge, and the pirates were the most democratic of all. Contemporary reports state that one man, who would later become famous for his exploits as a pirate captain, was most certainly there: the Chevalier de Grammont. De Grammont would eventually sail with some of the most famous men in the history of seventeenth-century piracy—Laurens de Graff, Jan "Yankey" Willems, John Coxon, and Thomas Paine. Given the long professional relationship that developed among them, it is likely that all or most of those men were together on Las Aves.

There is also circumstantial evidence to suggest that the most famous pirate of all, William Kidd, was one of the men of Las Aves. Kidd was serving as an officer aboard a French privateer ten years after the wreck at Las Aves, and he served in the English navy during the third of the Dutch Wars (1672–78). He may have joined the French in this venture in order to carry on the fight with the Dutch. He was certainly friendly with some of the buccaneers, most notably Thomas Paine, who sailed with d'Estrées.

These buccaneers were men who recognized opportunity when they saw it, and in this disaster, opportunity is what they saw. Why return empty-handed to Tortuga, after so many had come so far? For three weeks the buccaneers continued their revels at Las Aves, and then they were ready for more action.

The repercussions of this chance meeting on Las Aves would sound throughout the Caribbean, Europe, and America for the next forty years. The decisions made on that barren island would set off a chain of events that would usher in an age of piracy never seen in the world before or since.

The careers of men whose names would later become synonymous with piracy—Blackbeard, Black Sam Bellamy, Bartholomew Roberts, William "Billy One-Hand" Condon, Charles Vane—were born of this meeting, and those later pirates were but a single generation and a few degrees separated from the buccaneers on that sun-baked beach.

But of course, the men on Las Aves were thinking of plunder, of gold, of sacking the towns of their perennial enemy, Spain.

They were thinking of Maracaibo.

7

Over the Reef

I was not thinking of treasure when I asked the conch divers to show us where the "sewer pipes" were. I had no idea what to expect, no real sense of the magnitude of the disaster that had taken place on the reef. For all I knew, I was going to find piping for a sewage treatment plant.

One of the divers cheerfully volunteered to lead the way through the reef. His name was Angel. Angel was perhaps seventeen years old and had been diving at Aves since he was twelve. He had become a creature of the sea: tall and lean, with tremendous legs that propelled him through the water with the grace and speed of a dolphin. Angel came back with me to the boat.

Later that morning, when the entire team was assembled and ready to go, Angel led the way. He was wearing just a pair of tattered briefs and dime-store mask and fins. We followed in complete dive gear.

It was just as rough on the second day as it had been on the first, but this time we had Angel's local knowledge. While we had spent the whole day before banging against the reef, running into a brick wall over and over again, trying to find a way through, Angel knew where the door was. Or, more precisely, where to find a narrow passage leading *through* the reef, not over it.

This is not to say that it was easy. Among our group were several strong swimmers. Still, the seas over the reef were nearly too much for any of us. It was like being trapped in the spin cycle of a washing machine. Experience and physical strength let you stay on the reef a little longer before you were swept away, but that was all.

Foot by foot, we fought our way against the current and out toward the sea; one by one, we were peeled off by the current. Some lost their grip and were thrown backward into the staghorn and fire coral.

Max was another story. Once, when he was learning how to wind-surf, he hopped on his board and headed out to sea. He had not yet learned how to turn around, but he was having so much fun that he just kept on going until he lost sight of land. Eventually he managed to get his board going in the opposite direction and got home after dark.

Max took off after Angel without the slightest concern for the return trip. There was no way I was going to let Max go out there without me. If he was drowned, not only was I not going back to Cape Cod, but there were also serious bragging rights at stake.

Hand over hand, we pulled ourselves through the passage in the coral, trying to keep ourselves streamlined while hugging the bottom. The water was very shallow. If you lifted up at all, the current would

Max Kennedy with the reef in the background

peel you off the bottom and fling you backward into the staghorn. My mask was ripped off by the rushing waters so many times, I finally pulled it down over my neck.

We crawled along about fifty yards, our faces in the sand, glancing up to catch glimpses of Angel leading the way, unencumbered by diving equipment. Then, as if we had entered the stillness of a millpond, we slid over the edge of the reef onto a plateau at the edge of an abyss that seemed to drop away forever off the edge of the earth.

The change was breathtaking—and we had little breath left—from the spin cycle of a washer to a colossal aquarium teeming with schools of iridescent fish in a profusion of color that was so overwhelming I had to sit on the bottom for a moment to collect my senses.

Looking up, I saw the underside of the waves steamrolling harmlessly overhead. The calm was so peaceful, however, that I nearly forgot about the return passage and sat there for a moment like the Cowardly Lion in the poppy fields of Oz. At the surface, I could see Angel and Max. Angel began to wave and point down. Max was beneath the waves in a flash. He emerged, yelling with excitement, "Cannons! Cannons! Right under me!"

By now, I was nearly out of air.

The euphoria of the coral shelf quickly vanished, and I was suddenly hit with the grim prospect of getting back in one piece. But I had to see the cannon. Against both logic and instinct, I swam to Max, and, hoping I could squeeze a last breath of air from my tank, I descended for a final look.

I thought I saw a couple of crisscrossed cannon, but cannon submerged for three centuries become so encrusted that they are hard to spot at a glance. It takes a long hard look to be sure that what you are seeing really are cannon and not, say, sewer pipes. I wanted to see cannon. I thought what I was seeing were cannon, but there was no time to be sure.

Back to the surface; back to the lagoon. It was time to pay the fiddler for a dance I didn't want.

Angel led the way, though the passage through the reef was not a big help on our return to the lagoon. Working against the current on our way out was exhausting, but it was relatively safe since we were moving slowly into the flow of water. Coming back was something else again, however.

With his knowledge of the reef, and no cumbersome dive gear, Angel was able to simply glide over the top of the coral, but we could not do that. Once we committed ourselves to the current, we were its captives. I used what little air I had to go underwater and get whatever protection I could from the few feet of depth. But when that air was gone, it was up to Lady Luck to get me home again.

Indeed, there would be no problem getting back in. The hard part was getting back in alive. Over the reef we went. We didn't know where we were going in that white foaming chute of water, and there was virtually no way to slow down.

We tried to control our path as best we could, but there was almost nothing we could do. We were smashed into staghorn coral, dragged over fire coral, and bounced off the reef. The coral slashed wet suits and skin, bruising us as we bounced along.

And that is how we got back across the reef. My legs and ankles were badly lacerated. My forehead had been stabbed by a piece of staghorn coral. Max hadn't fared much better.

Michael Mailer, badly lacerated early on, and I went to the tiny coast guard station on the island for treatment. We had got the worst of the reef. But the reef failed to get the best of us.

That night, sleepless with my skin burning from fire coral, I wondered if what I saw were really cannon, or if my imagination had been playing tricks.

8

—

The Blue Lagoon

M ax and I returned from the windward side of the reef, looking as if we had been thrown through a plateglass window. Other than Angel, we were the only people who had made it over the reef. When we began to describe to the others what we had seen, Charles chimed in, "Ah, yes, I saw them, too, they were beautiful!"

Max and I looked at each other. Chris was with Charles when the current washed them out; after that they had spent the entire time together in the shallows looking for pottery shards. Chris confirmed this.

It was a minor triumph to have made it over the reef; no big deal, a few bragging rights, no more. But Charles had not been able to take even so small a one-upping. The former Olympic swimmer had not made it out of the lagoon, so he lied. Was this just an error in judgment, a silly fib to protect a fragile ego? Or was it a clue to the darker side of Charles Brewer-Carius? Indeed, I was reminded of the lesson James Bond learned when he caught Goldfinger cheating at golf, thus exposing the true character of Mr. Goldfinger.

Any vessel come to grief on Las Aves would, of course, have gone down on the seaward side of the reef. The next day, however, the surf was even worse, and there was no going over the reef for anyone.

Max and I had had a glimpse of what we had come looking for anyway. Although I was not sure that what I had seen were cannon, I was sure that they were concretions—encrusted masses of metal that could only have come from an old shipwreck.

We decided to look around in the relative shelter of the lagoon. We were rewarded almost immediately. Like Chris and Charles, we discovered heaps of smashed pottery. Only one pot was intact, but they all appeared to be of the same design. They had round bottoms and apparently had once been sealed on top—perhaps to hold liquids such as wine, olive oil, or even water. It was an exciting moment—the first tangible evidence that we might be onto something. It was unquestionably the detritus of a shipwreck, and an old one.

The shards were scattered everywhere. Unfortunately, many of them were in waist-deep water on the reef, so once again we were fighting the current. To keep in place, we would anchor the small boat and trail astern safety lines with large plastic balls fastened to the ends. If a diver "washed out," he could grab the line before being carried away by the current. And that is how we examined the evidence before us, looking at as many of the shards as we could. With every shard we found, it looked more and more as if we had stumbled on a major find. I began to think that we were dealing with more than one wreck.

At many wreck sites, most of the artifacts are in situ, remaining close to where they fell. And, by carefully noting their location on the bottom, one can get a good idea of how the ship was built, and how life aboard was organized. But that was not the case at Las Aves. It was clear that what we would find on the windward side of the reef would be heavy objects such as anchors, cannon, and ballast stones, the rocks stored at the bottom of a ship to keep it upright. Everything else would have been swept away. While wooden barrels would have drifted into shore relatively soon after the wreck, the pots, full of liquid, must have initially sunk straight to the bottom. Then, through the years, the current pushed them over the reef. With our experience so far of the currents, it was not hard to believe that the sea had carried the artifacts into the lagoon, breaking them on the way, just as we had nearly been broken.

We measured and photographed the single intact pot. Charles took the pot and said he was going to present it to the Ministry of Culture, although I argued with him to leave it where it lay—that it wasn't worth the risk of a smuggling charge if he were caught with it before he could get it to the ministry. But there was no arguing with Charles once he had made up his mind.

For the remaining two days of the trip, the wind never let up, and we never made it over the reef again. I took the opportunity to explore the island of Las Aves itself with a metal detector. Chris came along with me. Though we did not yet understand the scope of what had happened here, we did know that there had been castaways on the island, and we were curious to see if we could find any evidence.

Las Aves has attracted few settlers through the years. The only structure there is a small Venezuelan coast guard station at the south end of the island. It's an odd building, shaped like an igloo. A half-dozen coastguardsmen patrol the inshore waters in a large open boat. They and the conch divers are the only people who work on the island, and they are all transients. We would soon see why.

Chris Macort metal detecting on Las Aves

Coast Guard station, Las Aves. Note the trees bending in the wind.

From the boat, the island had not looked like a place where I would like to camp, but once ashore I realized just how brutal a place it truly was. The grass is short and stiff and makes walking barefoot painful. It is very hot, and there is no shade. There are no trees other than swamp mangroves. The place is infested with bugs and swarming with flies. The only ponds on the island are small salt ponds, and, when they dry, they look like the tortured surface of a distant planet.

We explored the northern tip of the island. We didn't find much of interest except for a series of stone-lined depressions in the ground. After a few sweeps with the metal detector with no results, we concluded that the depressions were old wells. We would later read in Dampier's *A New Voyage Round the World* that the wells had been dug by privateers who had put into the island to water their ships. They did not look as if they would have been able to supply any significant number of castaways, even in their best days.

We also found chunks of thick green glass from broken bottles. These were the heavy green "onion" bottles that were in common use in the late seventeenth century. So called because of their shape, there must have been hundreds of them. We wondered if perhaps this was a spot where survivors had sat to drink away their sorrows. I picked up the neck of a bottle, its fat lip still attached. I imagined the pirate or

Ancient well on Las Aves

sailor who last held it to his parched lips, and saw what for a thousand suffering souls had not changed in three centuries—or thousands of years before that.

We also found more of the same terra-cotta pottery we were finding on the reef. That was tantalizing evidence that we had found a campsite.

We continued to look around inside the lagoon, but we couldn't do much more than dive down and look at things and confirm our suspicions that we had found a wreck site.

Of course, that initial trip to Las Aves had not been undertaken as a major expedition, just some fun and occasionally life-threatening diving in the tropics. We poked around for four days, and then the expedition came to an end. It was time to go.

Leaving Las Aves by air is perhaps more dangerous than by sea.

A Cessna 210 was piloted by a local, who had made the trip many times before. He was arguing with the passengers about the extra baggage being loaded aboard as the wind lashed across the small grass strip. Tiny cyclones of dust whirled in the hot air.

The runway appeared short to me, especially considering the large earthen embankment at its end. I was debating whether I should insist that the plane not leave. But, as you often do in those situations, you simply pray.

The pilot took the Cessna to the far end of the runway—so far that her tail extended into the brush. He wanted every inch he could get for his takeoff.

There was not a trace of apprehension in the faces of Max and Pedro, who had "beeg Coobans" between their grinning teeth. The pilot revved the engine and popped the brake. But rather than take off like a hot rod, the little plane hesitated. Then, in the most undignified manner she began to waddle down the runway, wheels spread and engine grunting. The pilot had to use every speck of runway to get the speed he needed to get his ship airborne, and a wheel kissed the embankment, leaving a puff of dust as they flew off.

I am happy to report that our departure was somewhat less dramatic.

Back in Caracas, Charles Brewer called a press conference to announce the extraordinary find we had made on the reefs. The conference was held at an exclusive country club, of which he and his family were members. The club was original Spanish Colonial, at least three hundred years old, with a panoramic view of Caracas.

Among the press corps in attendance was the Associated Press correspondent in Venezuela, Bart Jones, an American. I was tired and ready to go back home. Charles had set up an interview for me with Bart, however, and he insisted I do it, so I relented.

Bart was wary. He had a poor opinion of Charles—especially his reputation for manipulating the press. He assumed that I was part of the Brewer publicity machine, and he was prepared to inhale a lot of ether.

Bart and I talked for some time, and, the more we talked, the more he realized that I was *not* going to hype the find as a treasure wreck as Charles intended to do. Finally, Bart asked, "What do you know about Charles Brewer?"

I equivocated a bit, suppressing an impulse to tell Bart about our experience on the reef with Charles. I told him I didn't really know much, which was true.

Bart knew quite a lot about Charles Brewer, and he told me a few things—most of which Charles had *not* mentioned in his résumé.[1] Charles Brewer was more well known in Venezuela than I had imagined. He had been Minister of Youth, and he had been a partner of the controversial anthropologist Napoleon Chagnon. Brewer and Chagnon had worked with the mistress of impeached Venezuelan president Andrés Pérez in an effort to take control of huge tracts of

Barry Clifford, AP reporter Bart Jones, and Charles Brewer discussing
d'Estrées' map

land on which the primitive Yanomami tribe live. According to Bart,
Charles had been caught in illegal gold strip-mining ventures.

From what Bart was telling me, I could see that, at the very least,
Charles was playing the same game here, working himself into the
center of things so that he could exploit a situation. For a guy who was
essentially a sports diver, he was already posturing himself as a great
maritime explorer and underwater archaeologist. Even before I left
Venezuela, I knew that if I had any interest in further exploration at
Las Aves, Brewer would make sure that he was going to be in control.

I wasn't so sure about some of Bart's other allegations. I had not
known him long enough to tell what his particular biases might be,
and some of his information simply staggered the imagination. It
almost sounded as if one of the early Spanish conquistadors had been
somehow reincarnated at the very brink of the second millennium. At
the back of my mind was also the Elizabethan river wreck of which I
had heard. I know the sea; someone like Charles would be needed for
a jungle river expedition. I decided to be wary of Charles, but to keep
an open mind.

Bart wrote the Las Aves story for AP, and it went out over the wire.
Soon the whole world was aware of what we had found at Las Aves.

Those stories were read with great interest in certain treasure-hunting circles.

Just a few days after the interview, I was back at my home in Provincetown, at the tip of Cape Cod. Looking out over the frigid, blue-gray Atlantic, as an icy wind kicked up rows of whitecaps, it was hard to believe that this was the same ocean in which I had been diving just a week before.

What we had found continued to tantalize me. There are, of course, thousands of wrecks scattered across the ocean bottom. Most are not worth the trouble to find. But Las Aves seemed to hold promise. I wanted to know more.

I gave Ken Kinkor a full account of what we had found on the reefs and asked him to look into what ship, or ships, might have been there. With the work on the *Whydah* wrapped up for the season, Ken had time to dig deeper into the history of Las Aves. What he found fascinated us.

Ken unearthed primary source documents, reports, and letters from English sources describing the magnitude of the disaster that had befallen d'Estrées' fleet, and the catastrophe it represented to French designs on the Caribbean. In the course of phone conversations with other historians and archaeologists, I gathered more material. Others who had seen the AP reports chimed in with what they knew.

The most important discovery was a map d'Estrées had made of the wreck site before departing Las Aves. It was an incredible document. The admiral's drawing of the reef system looked very much like modern charts of the area. All along the reef line were drawings of French men-of-war positioned at the places where they had struck. Next to each of the carefully drawn pictures was the name of that unfortunate vessel—except for two. Next to those drawings was only a single word: *flibustier*.

Flibustier. A French word derived from the Dutch *vrijbuiter*; in English, it is "filibuster." All are terms for the same root concept, "freebooter," defined by the *Oxford English Dictionary* as "one of a class of piratical adventurers who pillaged the Spanish colonies in the West Indies during the 17th century."

With that one word, my interest in the wrecks at Las Aves skyrocketed. *Filibusters.* Pirates have always been my main area of interest. At that time, the *Whydah* was the only known pirate shipwreck ever discovered and authenticated. Now, here were the locations of two more. Not eighteenth-century pirates like Bellamy, but seventeenth-century buccaneers of the Spanish Main.

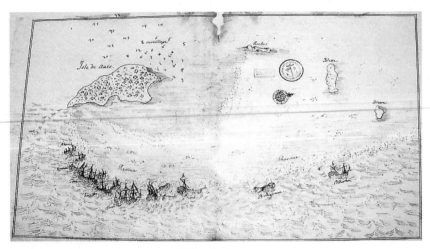

D'Estrées' map

More research revealed the French recruitment of the buccaneers of Tortuga, at least fourteen hundred pirates on fifteen ships. Knowing that Tortuga was the central gathering point for the Caribbean pirates of that era, and knowing that the French must have recruited nearly every major crew in the West Indies to have put together a flotilla of that size, Ken suspected that some of the famous figures in the annals of seventeenth-century piracy might well have been among those men of Las Aves.

It was an extraordinary prospect. D'Estrées had lost some of the largest warships of his time on the reefs at Las Aves. The buccaneer contingent represented one of the largest mobilizations of those men ever recorded. And here were to be found the remains of ships from that event, artifacts from the beginning of a great wave of piracy that would plague the Caribbean for decades.

I was ready to go back to Las Aves.

9

The Chevalier de Grammont

[T]he chief of the filibusters.
—*THE SCOURGE OF THE INDIES*
Maurice Besson

The Brethren of the Coast were not men who took orders easily. Power and influence could shift radically and quickly within their community. Leaders emerged, had their moment of power, and then fell victim to any of the many fates that awaited such men—death in battle, shipwreck, or at the end of a noose, their men turning on them, losing face for a bad decision or a momentary act of perceived cowardice. Many buccaneers rose to power and fell. At the time of the Las Aves disaster there was only one undisputed leader of the filibusters: the Chevalier de Grammont.

One of the most popular Hollywood pirate themes is the banished nobleman turned pirate, the aristocratic gentleman forced to flee his ancestral home and turn buccaneer. Such was the case with de Grammont.

The Chevalier was a small, swarthy man in his late twenties or early thirties at the time of Las Aves. His father, who died when de Gram-

The Chevalier de Grammont

mont was quite young, had served in the King's Guards. Though the family was not of the uppermost strata of French society, they did enjoy a certain status and the favor of the royal court. As it turned out, that was fortunate for the Chevalier.

Little is known of de Grammont's earlier years. Indeed, even his Christian name is in some doubt, given variously as Michel, Nicolas, or François. He was born some time before 1643, during the reign of Louis XIII. The story (perhaps "legend" is a more apt term) of his road to piracy is the perfect pedigree for the swashbucklers of fiction.

From an early age, de Grammont displayed all of the pride that came with being a member of French nobility. With his father gone, the young man considered himself head of the family, even if others did not necessarily consider him so.

When de Grammont was in his early teens, his mother remarried. Her new husband, like de Grammont's father, was a military officer.

At some point after becoming de Grammont's stepfather, he introduced de Grammont's sister to a fellow officer, who he thought might make a suitable match.

The young Chevalier did not agree. He felt his sister's suitor was below the family's station, and he made that opinion well known. His sister's opinion of the suitor is unknown, but was probably considered inconsequential. Marriages among the French nobility were made on the basis of considerations other than those of the heart.

To de Grammont, it was a matter of family honor. Once, when the Chevalier's stepfather was not at home and the suitor came to call, de Grammont had the servants forcibly eject the man.

Despite this insult, the suitor continued his courtship and continued to treat de Grammont like a child, making light of the Chevalier's objections. To be thus dismissed must have been infuriating to the proud young man.

It came to a head at last when de Grammont informed the officer that if he were a little older, they would cross swords. Far from being intimidated, the officer continued to mock the Chevalier until, in a fit of rage, de Grammont snatched up a sword and went for his tormentor.

The young officer, not wishing to hurt his beloved's brother, did no more than fend off the attack, but de Grammont was out for blood. Twice he managed to wound his adversary. Thrown off by the wounds and the intensity of de Grammont's assault, the suitor failed to turn the final thrust aside. The Chevalier de Grammont delivered a fatal wound.

Fatal, but not immediately so. De Grammont's servants carried the dying officer away to his house, where he lingered on for two more days. The king sent a major of the Guards to visit the man to determine what had happened and who was guilty of this crime.

Generously, the wounded officer explained that the fault was his, that he had provoked the affair and that it had been carried out with honor. Even more impressive, he sent the Chevalier de Grammont enough money for him to escape France rather than be tried for murder. Finally, he bequeathed to de Grammont's sister, whom he would not live to marry, the sum of ten thousand livres.

As it happened, between the officer's deposition and the influence that the de Grammont family enjoyed in court, the Chevalier escaped banishment and complete disgrace. With the scandal hanging over his head, however, it was thought advisable for de Grammont to absent

himself from Paris, so he was given a commission in the Marine Regiment. In that service de Grammont first saw the West Indies.

The Chevalier de Grammont was bold and fearless, and he served
with distinction. After several years, he was given command of a frigate.
Near Martinique he captured a Dutch convoy worth more than
400,000 livres. The prizes were taken to the French colony in Saint-
Domingue (Haiti). De Grammont had to give up a goodly portion of
the prize money to the king, but he did receive one-fifth, an enormous sum, as a first taste of the possibilities that sea robbery offered.

Pirates were men apart from society. As such, many chose to engage
in excess in all aspects of their lives. Though de Grammont was a reliable and capable officer, he seems also to have been a man who
enjoyed a good time. Just eight days after coming ashore with his prize
money, de Grammont found himself once again at sea, having in that
short time blown nearly all of his newfound fortune on gambling,
prostitutes, and other debauchery. Though he was still in the king's
service, he was already acting more like a buccaneer than a marine
officer.

De Grammont's second cruise for France was not so successful.
Before taking any prizes, his ship was driven by storm onto a reef,
where she broke up. It was then that the Chevalier turned to piracy.

FIRST AMONG EQUALS

Why, exactly, this member of French society, this well-respected officer, turned to buccaneering is not quite clear. There is no indication
that he was out of favor with the military, or that the loss of his ship
would end his career. Perhaps de Grammont did not care to labor
under the eye of superior officers. Perhaps his brief taste of wealth
from the Dutch convoy had whetted his appetite for more, and he did
not care to share any further good fortune with the king. Whatever his
reasons, the Chevalier de Grammont turned pirate, and it proved to be
a job for which he had a natural talent, and one in which he rose
quickly among his brethren to command.

The pirates of the later seventeenth century were not looked upon
in the same way as were the pirates of the early 1700s—Blackbeard,
Bartholomew Roberts, and their kind—who were despised as outlaws
and the great villains of the age. In the Chevalier's time, the situation

was somewhat different. Pirates of the late 1600s were often called "privateers," meaning that they actually had legal authority to plunder the wealth of enemy nations—a license to steal.

Such distinctions often became a bit hazy. A good deal of what these "privateers" did was genuine piracy. Still, the buccaneers were not shunned as the later pirates were, and freebooting could even be a stepping-stone to respectability and power.

The early buccaneers may have called themselves "the Brethren of the Coast" as a rejection of the authority of formal government, but they still held national and religious loyalties that would be rejected by pirates four decades later.

The later pirates declared war on the whole world, but the early buccaneers focused on the Spaniards. For pirates of English descent and the French Huguenots, the fight was in part religious. Catholicism infused every part of the Spanish national character, and that was unacceptable to most Englishmen, who would soon exile James II, their last overtly Catholic monarch.

For the French, it was a matter of Spain's attempts to dominate the West Indies. Spain at this time was like the USSR after Afghanistan, a superpower on the decline, a shell of what it had been, but still trying to exert its influence, still trying to keep the Caribbean a Spanish lake. That attitude was resented, and it was becoming untenable.

For these reasons, it was easy for the legitimate governments of England, France, and Holland to tolerate the buccaneers and, indeed, to recruit them when needed.

Having brought his captured Dutch convoy to Saint-Domingue, de Grammont was well respected by the filibusters there, who were apt to be impressed by such things. From Saint-Domingue, the Chevalier proceeded to Tortuga, the epicenter of pirate activity in the West Indies. There, with the last of his money, he fitted out a fifty-gun ship.

The Chevalier de Grammont, a natural leader, quickly rose to prominence among the Brethren of the Coast. As a modern writer would put it, "grace, eloquence, a sense of justice and a distinguished courage soon caused him to be regarded as the chief of the filibusters."[1] Buccaneers swarmed to him, eager to take part in any exploit he had in mind.

And so, in April 1678, when Governor M. de Pouançay, acting on the orders of King Louis XIV, called for the buccaneers of Tortuga to join in an expedition against the Dutch at Curaçao, de Grammont found himself de facto leader of the filibuster contingent. When they found themselves thrown up on the sands of Las Aves, their plans of sacking Curaçao as shattered as the French fleet, the Brethren of the Coast looked again to de Grammont for leadership.

No doubt it was to de Grammont that d'Estrées appealed in hope that his fellow Frenchman would persuade the pirates to continue with the attack on the Dutch outpost.

One has to wonder what these two men must have thought of each other. D'Estrées was a generation older than de Grammont and from a somewhat better family, but still they were both from the upper strata of French society. Did d'Estrées look with disdain on the Chevalier, who had brought scandal to his family and had left the honorable service of the king to turn pirate? Did de Grammont sneer at d'Estrées as a king's toady who could not even keep his fleet intact?

Certainly d'Estrées could not have been pleased with de Grammont's men's looting of whatever washed up from the wrecks, and must have resented the Chevalier's mercenary attitude toward an attack for the greater glory of France. For his part, de Grammont was not impressed by d'Estrées' exhortations to continue on to Curaçao and clearly thought that the admiral had no realistic chance of taking the island.

These two sons of the French nobility faced off under the blazing Caribbean sun, on a desolate beach amid the ruins of the French fleet and surrounded by drunken, dispirited, angry, and dying men. The disgraced nobleman-turned-pirate held all the cards in those circumstances, which must have been the greatest irritant of all to d'Estrées. In the end, de Grammont and the filibusters would not be moved to continue in the service of France, and d'Estrées could do no more than sail away.

The Brethren of the Coast reveled for a time, enjoying the unintentional largesse of the French, until it was time to move on. No doubt there was a lively discussion as to where they should proceed, with nearly all of the Spanish Main under their lee. Though de Grammont, with his natural flair and qualities of command, was looked upon as the leader of the expedition, there was most likely a vote as well. That was the way of the buccaneers. They had not escaped European tyranny just to impose it on themselves in the Caribbean.

10

The Sack of Maracaibo

Thence we sailed against the Spaniard with his hoards
of plate and gold,
Which he wrung with cruel tortures from Indian folk of old;
Likewise the merchant captains, with hearts as hard as stone,
Who flog men and keelhaul them, and starve them to the bone.
— *"THE LAST BUCCANEER"*
Charles Kingsley

JUNE 1678
VENEZUELA

Buccaneering in the seventeenth century consisted primarily of land-based raids. Treasure ships were hard to find on the great expanse of ocean. They sailed in large convoys that were hard to attack, and these "plate fleets" did not keep rigorous schedules, making it difficult to know when and where they might sail. On the other hand, the cities and towns where Spanish gold and silver were warehoused did not move. These cities contained wealth far beyond that intended for the royal coffers. It is likely that the men of Las Aves immediately focused on plans for the sack of some Spanish colonial town.

The pirates settled on an attack on Maracaibo, Venezuela, about four hundred miles to the west. This was a fair-sized Spanish city situated on the west side of a narrow channel through which Lake Maracaibo emptied into the Gulf of Venezuela, known to the pirates as the Bay of Maracaibo.

Alexandre Exquemelin, a buccaneer himself, two decades earlier described Maracaibo as

> very pleasant to the view, by reason its houses are built along the shore, having delicate prospects everywhere round about. Here also one Parish Church, of very good fabric, well adorned, four monasteries and one hospital. The inhabitants possess great numbers of cattle and many plantations, which extend for the space of thirty leagues within the country, especially on that side that looks toward the great and populous town of Gibraltar.[1]

Though a raid on Maracaibo was potentially profitable, it was hardly an original plan. Henry Morgan and François L'Ollonais had both sacked the town years before. Of course, in the waning days of Spanish rule in the Caribbean, it was hard to find a city or town of any importance that had not been sacked at one time or another.

In early June 1678, de Grammont and his fleet of six large ships and thirteen smaller ones, manned by well over one thousand men, sailed from Las Aves to Maracaibo. De Grammont put half his force ashore, marching them along the San Carlos peninsula, intending a landward assault of the fortification there, a fortress aimed at defending against attack from the sea.

The Spanish troops garrisoned at Maracaibo held the filibusters off for a little while, but de Grammont managed to land heavy guns and bring them to bear on the fortress. It took no more than a brief cannonade to convince the Spanish that their situation was untenable. They surrendered the fort to the buccaneers. With the heavy guns of the Spaniards secured, de Grammont then took half the buccaneer fleet over the shallow bar and left the rest to blockade the approaches to the city.

Maracaibo flew into a panic. Those who could, including the new governor, Jorge Madureira Ferreira, abandoned the city and raced to the relative safety of the countryside or neighboring towns. De Grammont took the city with virtually no opposition, and he and his men set to plundering it with a will.

By the latter half of the seventeenth century, the Spanish cities of the Caribbean had been attacked again and again by sundry sea raiders in the same way that the towns of England had once been plundered repeatedly by Vikings. The Spanish people in that part of the world, like the English victims of the Vikings, had come to expect the most vicious kind of brutality: murder, rape, arson, torture. They had no reason to expect anything less from these new seaborne attackers. After all, it was precisely the same treatment they meted out to the Native Americans.

The pirates from Las Aves were brutal and efficient. Some went after the governor and other refugees who had managed to escape, chasing them farther into the back country. The others robbed the city of everything they could find. Then they began to torture the citizens to discover if there was anything else secreted away.

Once they had finished with Maracaibo, a contingent of the buccaneers crossed to the eastern shore and fell on the city of Gibraltar. Again, it took only a short bombardment to induce the twenty-two Spanish soldiers defending the city to give it up to the pirates. Gibraltar, unlike Maracaibo, had not been taken by surprise. Having had the benefit of the two weeks the buccaneers had spent sacking Maracaibo, the citizens of Gibraltar had packed their valuables and abandoned the town.

The buccaneers did what they always did—they followed the money. De Grammont marched his men inland fifty miles for the town of Trujillo, to which the refugees of Gibraltar had fled. This was no easy stroll, but rather fifty hard miles through the deadly pestilence of the South American jungle, one of the reasons the buccaneers preferred to strike from the sea. Many of de Grammont's men died along the way.

At last they reached Trujillo, which was defended by a fort sporting four artillery pieces and 350 Spanish soldiers. Exhausted from the march, outnumbered, and vastly outgunned, the filibusters nonetheless attacked.

De Grammont and his band of pirates stormed the fortification, coming at it from the rear, as one defender put it, "by some hills where it seemed impossible to do so." The Spanish again fled before this invading force, refugees struggling to get to the next town, Mérida de la Grita, seventy-five miles away.

A study of such buccaneer raids reveals how often the pirates won against overwhelming odds in the most unlikely situations. Greatly outnumbered by regular Spanish troops, desperate and beyond caring, the pirates threw caution to the wind and attacked. Despite the odds against them, their attack succeeded.

No doubt this was due in part to the ferocity and determination of the pirates, and perhaps to a lack of motivation among the Spanish troops, who sensibly chose to run for their lives rather than die defending the wealth of the Spanish aristocracy. The buccaneers' unorthodox tactics played a role as well. While regular troops were accustomed to fighting European-style battles, the pirates employed guerrilla-style tactics, making it hard for defenders to predict when, where, or how they would attack. Further, the buccaneers, many of them former hunters from Hispaniola, were better marksmen than the Spanish troops. Whatever the reason, the pirates continued to succeed, time and again, against overwhelming odds.

De Grammont had not only taken a succession of Spanish towns but in fact had made himself master of the entire Lake Maracaibo region, with no local force to challenge his supremacy and no need to rush his looting. After taking what they could from Trujillo, de Grammont's men returned the way they had come, once again occupying Gibraltar. For a week or more, they continued to plunder the town. When they had taken everything they could lay their hands on, they burned it.

In all, de Grammont spent nearly half a year on Lake Maracaibo, raiding, debauching, looting, and burning. It was not until December 3, 1678, that the Chevalier and his fleet, heavy-laden with all the wealth wrung out of the Lake Maracaibo region, left the Gulf of Venezuela.

They did not return to Tortuga. Rather, they made for Petit Goâve, a hell town in Hispaniola, fifty miles west of present-day Port-au-Prince. Petit Goâve was beginning to challenge Tortuga as the chief gathering spot for the buccaneers. De Grammont and his men arrived as heroes.

It was no matter that the war in Europe, which had been the root cause for collecting together the buccaneers in the first place, was winding down. The pirates were barely interested in such formalities. Their hatred of Spain and their disdain for treaties between nations went far deeper than that. They had, in fact, only just begun.

Vast armadas of buccaneers were not a new thing in the region. L'Ollonais, Morgan, and others had already used that weapon as a tool of colonial policy. But something had begun on those hot sands of Las Aves that would not easily be stopped. With no sort of legal authority, the most charismatic leaders of the buccaneer community had come together, had led a great army in half a year's raid on Spanish settlements, and had come away rich for their efforts.

The French had brought the buccaneers together. The destruction of the fleet at Las Aves had ended the mission for which they had organized. Rather than return to port, however, the pirates had stuck together and had launched a raid of their own choosing.

With the raid at Maracaibo they had set a new precedent, formed a loose alliance, an army that would split and come together again at will, like quicksilver. These buccaneers would become the dominant force in the Caribbean, and remain so for years to come. Governments, despairing of stopping them, would instead try to lure them into their service.

The wreckage of the French fleet on Las Aves, as it turned out, was the starting point for some of the greatest piratical careers in Western history.

De Grammont's raids after the attack on Maracaibo

11

Curiosity Sparks Expedition

The more Ken and I dug into the history of the wrecks on Las Aves, the more we understood how the disaster had kicked off a wave of buccaneering from which emerged some of the leading names among the seventeenth-century pirates, and the more eager I became to get back there. During our brief stay on the island, I had seen enough to convince me there was a mystery to be solved at this site. What had the filibusters left behind? What couldn't they find? That was one thing that interested me. Given the magnitude of the disaster, it seemed impossible to me that a longer, more systematic search would not reveal much, much more.

The second thing that interested me was the map d'Estrées had made. The admiral, intending to go back one day and salvage what he could of the wrecks, had made a meticulous drawing of the island, the reefs, and the ships that went up on them.

I was curious to see how accurate d'Estrées had been. If his map was indeed accurate, then it would be a useful tool for discovering wrecks. Sometimes you get lucky, but more often you have to prod your luck along though tremendous research and hard work. We had labored over maps and other documents to find the *Whydah,* but it all turned out to be right there, waiting to be unearthed. I have found other wrecks as well by finding documents and following their lead. Perhaps d'Estrées' map would be one of those.

Of course, I was excited by the "filibusters'" ships. Pirate ships generally disappeared with little fanfare. Clever pirates, like Thomas Paine, did not keep records. But in my hands was an official map, made by an admiral of the French navy, noting the location of two of them. It was a chance to be the first person in three hundred years to set eyes on a pirate vessel of the Spanish Main, to see how it might differ from the *Whydah,* a pirate ship of forty years later. And I wanted to be the first to swim, see, and touch another pirate ship.

I called Mike Quatrone of the Discovery Channel, who has the nose of a truffle hound when it comes to sniffing out a good story. We discussed organizing a full-scale survey and exploration of the reefs at Las Aves, of going back and shooting a documentary about the wrecks. Soon our talk evolved into a plan: with the Discovery Channel, in tandem with the BBC, underwriting the exploration and producing a show about the expedition.

With some of the other people involved in the project, Pedro Mezquita and Charles Brewer and their contacts in Venezuela, we talked about a major excavation and a museum in Caracas dedicated to the wrecks of d'Estrées' fleet. We began to develop plans for a conservation center and discussed training Venezuelans as conservators. With high hopes and the best intentions we envisioned an important project.

Charles took the lead in Caracas, talking to everyone he knew, which seemed to be a lot of people. I was afraid he was overstepping his bounds, but the phone calls and e-mails I received from him seemed to indicate that we were all on the same page. As we were making preparations, Charles (who writes English exactly as he speaks it) sent me an e-mail, addressed to "Dear friend Barry," discussing how he was allaying some of the Venezuelans' fears about what we intended. He wrote, "I am dealing here with people that are full of fears and complexes of underdevelopment (mentally). Questions as Who are these guys?, Are they coming here to smuggle items? Should we put the police to follow them?" He closed by assuring me that he would be "solving problems" for me in Venezuela.

It was an awkward situation. The BBC and the Discovery Channel had agreed to film a documentary about the exploration based on our work with them in Scotland on the King Charles I site. But Charles seemed to have contacts within the Venezuelan government. Still, I wondered if I was being given "the ether."

We were both eager to get back to Las Aves to see what we could find. We had to figure out how to work with each other.

Then, from left field, came another problem. A high-powered insurance salesman from Florida had invested money with a fly-by-night Florida treasure hunter. The initial ill-conceived project was to locate a sunken German U-boat off Venezuela, allegedly filled with stolen Jewish gold. One of the investors' former employees told me that $600,000 had been put up but nothing was ever found: no U-boat and certainly no gold. The treasure hunters needed a big find to keep their investors on the hook.

The Florida businessman had two Venezuelan partners with a lot of influence. At some point prior to our going down to Las Aves they secured a contract with the Venezuelan navy to do "archaeological" salvage work in that country. They claimed they had only to ask for permission and they would be allowed to conduct underwater archaeological digs wherever they wanted.

When one of the Florida treasure hunters read Bart Jones's article in the newspaper, he fired off an angry letter to me, informing me that I had no right to explore their wrecks, that Las Aves was their permit area, and that the right to explore Las Aves was exclusively theirs. He claimed that the Venezuelan navy, having known about the wrecks at Las Aves, had asked that his team go out there and conduct archaeological work.

To this day, I still don't know how valid those claims were. I certainly wasn't going to jump someone's permit. I have been on the receiving end of that game too many times, with the *Whydah* and other projects. To complicate matters, we had also heard rumors that a notorious wrecker had been bragging that he had taken a cannon from the site and was making plans to do his own salvage, with or without a permit. Business as usual on the treasure-hunting front.

The question was, could we legally document these sites before they were lost forever? In a place like Venezuela, when one government agency gives you "exclusive" rights it doesn't necessarily mean it has the authority to do so. On the one hand, I had this guy claiming exclusivity, but I also had Charles Brewer assuring us that he had all of the necessary permits in order for us to return to the reefs. After considering it all, I decided that we had as much right as the Florida group to work Las Aves. After all, we were only going down to map and film the site, not excavate. We decided to press on.

An interesting sidebar to all of this was the reaction of the archaeologist Dr. John de Bry, whom the Florida group had hired. De Bry, who grew up in France, is an expert at researching historical documents. At the urging of the group's backer, he did a great deal of work on the Las Aves site. He even spent time at Las Aves, several months after our visit, exploring the wreck sites. He ended up leaving the project when he discovered that his employers were more interested in treasure hunting than archaeology.

John de Bry had heard plenty of stories about me from the people on his Las Aves venture, and his impression was that I was Genghis Khan in a wet suit, determined to loot every site I could find. So it didn't sit very well with him when a friend told him to watch a report on CNN about our Las Aves explorations and he saw, on television, his own research material in my hands!

Maps and other archival material are public domain, but taking research that someone else has done is theft of intellectual property. De Bry called his lawyer and had him write to me, demanding that I stop using his material.

I certainly understood that. Copies of the material had been sent to us by way of the *Boston Herald*. It turned out that these copies had been sent to the *Herald* by a diver who was a disgruntled employee of the Florida group, and who, as part of the team, had been given copies of de Bry's work. I wrote to John's lawyer, explaining this. But the fact is, I had previously found a copy of the same map de Bry had located, independent of the *Herald* copy. I found mine published in a French history book. I sent a photocopy of the page from the book to de Bry's lawyer with an admittedly sarcastic note attached. Then I braced for a good fight. Knowing the people with whom de Bry was associated, I looked forward to it.

More letters went back and forth, and as John read them he started to sense that perhaps I was not just an unscrupulous treasure hunter. My side of the story as to our research made sense to him. So one day, out of the blue, he called. We ended up talking for some time, since we obviously have a lot of mutual interests, and we developed a friendship over the phone. The National Geographic Society was staging an exhibit of *Whydah* artifacts in Washington, D.C. John mentioned that he would like to see it and I invited him.

A year or two later, when I needed an archaeologist for a project I was doing in Africa, I gave him a call. We have worked together on several expeditions now, and have become friends. John is from the old school, a Vietnam vet, and has the work ethic of a badger.

The planning, organization, and logistics of an expedition like this are complicated and difficult. We worked hard at our end getting the people and equipment together for the dive. Max and his friends were planning to be there for part of it. But I knew that the second trip was not going to be a clambake and Max knew it. He was going to have his hands full keeping his group in one piece. This would not be a vacation.

12

Logistics

Nearly all of the Las Aves team members were veterans of Expedition *Whydah*.

Even if Todd Murphy and I hadn't been friends for as long as we have, he would still be a man to have on an expedition like this. Todd is a U.S. Army Special Forces master sergeant (MSG) with twenty-three years of active duty and reserve time. He is currently in a National Guard Special Forces unit. Todd is trained as a combat diver, diving supervisor, and diving medical technician. He is also an instructor at the Special Forces scuba school in Key West, Florida.

Todd and I began diving together long before the *Whydah* project. We met on Martha's Vineyard in the mid-seventies. I had built a little Cape, with a great central fireplace, where, on long winter nights, I'd sit by the fire with my kids, reading the history of lost ships and sunken treasure. Todd would take my kids swimming and hiking and spend evenings with us.

I ran a small salvage/diving company at the time and was engaged mostly in doing emergency dive work for the Coast Guard, disentangling nets and cables from the wheels of fishing boats, and, on occasion, landing lucrative salvage contracts.

John Kennedy Jr. worked for me then as a diver. Subsequently, he and Todd became friends. The two of them were always looking for any excuse to go diving with me for shipwrecks around the Vineyard and Elizabeth Islands or to investigate various Island legends, such as the story of the two bronze cannons spotted by Alfred Vanderhoop lying off Gay Head, or the ancient Spanish helmets found in Quista Pond, or the lost cargo of lignum vitae logs resting somewhere on the bottom of Vineyard Sound. One trip that we made over and over again was to No Man's Land to find Viking Rock, a large boulder, reportedly with runic carvings, which was toppled into the ocean during the 1938 hurricane. Another of our hunts was for the wreck of the *John Dwight*—a ghost ship from the rum-running years, and the scene of the Vineyard's greatest murder mystery.

I think we explored every inch of every sandbar and reef, amassing an impressive collection of old bottles and china.

Todd and John were eager to go on these adventures and to pick up a few dollars on salvage jobs when they could. Todd was paying his way through college, and John was always happy to get forty dollars for a hard day's work. It meant a lot to him to earn his own money—that's one of the things I liked about him.

We considered each job as a dangerous rival that had to be carefully assessed before going forward. These were wonderful days, and I remember them as some of the best times of my life.

That was about the time I became fixated on the *Whydah*. Not surprisingly, so did Todd and John. John made the first dive for the *Whydah* at Marconi Beach on a cold, blustery day in November 1982. The magnetometer put us over the wreck, but what we hadn't factored into the equation was that there was twenty feet of sand that had to be dug through to get at it.

We came back in the spring of 1983 for a full-scale search. Todd and John were members of that first exploratory team. It was not as glamorous as they imagined. Their first job was to head up to Maine to work on our newly purchased salvage boat, the *Vast Explorer II*.

Todd was somewhat inexperienced when we began diving in the summer of 1983. However, after a few years of experience, he earned the job of diving supervisor/director of operations. He has been working with me ever since.

John went off to New York, though he often stopped by the *Whydah* HQ and museum at the end of Macmillan Wharf. The last time I saw John was in the spring of 1999. He had come to Provincetown to spend the day with Captain Stretch Gray and myself aboard the *Vast Explorer*.

Later that day, I found him wandering alone through the *Whydah* museum. The museum was closed for the season, and the cold Atlantic chilled the old building through its timbers. A smell familiar to those who have worked ancient shipwrecks permeates the air. It is the smell of pine tar and hemp, concreted cannon and flintlocks, and the belongings of long-dead pirates. It is more than a smell; it is an awareness that seems to drift about the place like a lost spirit. Perhaps it is.

Nonetheless, John chuckled out loud, remembering faces from old photographs of project crewmen that hang on the wall. Faces from when the thrill of finding pirate treasure was merely a dream—until it came true.

Todd Murphy keeps a cold-weather bag and another for the tropics ready at all times so he can grab the appropriate one and go if a project comes up. For example, it was less than two weeks from the moment I received a frantic call from a particularly mad Scotsman asking me to search for the wreck of *The Blessing of Burntisland* of King Charles I to the time we launched our first dive with full surface support.

By an odd coincidence, Todd was in Haiti with Special Forces about the time I was having troubles with the competing group at Las Aves. He happened to meet Dr. John de Bry, the former archaeologist for that group, who was then doing archaeological work at Cap Haitien. They started discussing their mutual interests, and we ended up hiring John for a project in Africa.

For the Las Aves expedition, Todd would be the director of operations, responsible for the thousand-and-one details involved in completing a successful mission. He would be assisted by Chris Macort.

Cathrine Harker is an archaeologist who had also worked on the *Whydah* project. She is Scottish, with a degree in geology from Edinburgh University and a master's degree in archaeology from Liverpool University. After graduating from Liverpool, she moved back to Edinburgh and became an exhibit interpreter for a traveling display of *Whydah* artifacts we had in Scotland at the time. That was where I first met her.

Since I needed staff for the permanent museum we were then

building in Provincetown, I asked her to join our team. Now, she would come to Las Aves to help explore and map the wrecks on that lovely reef.

Cathrine is very capable, and she is, without a doubt, one of the toughest and most loyal people on the team. More than once, I've seen her stand as solid as a samson post aboard the *Vast Explorer,* with breakers coming over the rail.

If Cathrine is feeling chatty, we might speak ten words to each other over the course of a summer. I think the English army understood it best: there are only two living creatures that will go face first into a badger hole—a Jack Russell terrier and a Scotsman.

Eric Scharmer was another diver who had worked on the *Whydah* project. An exceptional athlete, Eric is a former member of the Pro Mogul Ski Tour. He had taken up film and video-camera work as a profession. I persuaded the BBC to hire him to be the underwater cameraman at Las Aves because he knew our dive system. He ended up being the perfect man for the job. Indeed, if you've ever wondered "how on earth did they get *that* on film?" it's men like Eric who do it.

My future fiancée, Margot Hathaway, would also be coming with us. Margot is a strong swimmer and diver and also an accomplished fine arts and still photographer who has worked with me on the *Whydah* site as well as three trips to Africa.

Carl Tiska was the only team member who had not worked on the *Whydah,* but his credentials were nonetheless impressive. He is a lieutenant commander in the Navy SEALs. As with Todd, Carl's military training was a great asset to the project. Even though our expeditions are more relaxed than a military operation, there is still a chain of command and an organization not unlike the military's.

Though I had yet to meet them, I knew there would also be the production team from the BBC, and Max's group. I had no doubt that Charles would be bringing some of his own people. It was shaping up to be a large crew, and, as far as the people I could vouch for were concerned, a good one.

About a month prior to the start of the expedition, I had a meeting in my home above the *Whydah* museum with executives from the Discovery Channel and the BBC to nail down some of the details about the documentary, plan logistics, and outline what we all hoped to accomplish.

The fact is that the Discovery Channel and BBC were taking a big risk with this production. All we had found so far were a couple of

Left to right: Chris Macort, Carl Tiska, our driver, Barry Clifford, and Eric Scharmer

cannons and some pottery. There was no guarantee that we would find more, though what we had learned through researching the wreck of the French fleet suggested we would. Still, the Discovery Channel and BBC had only my word to go on. They were familiar with the *Whydah* project and what we were accomplishing, and so the consensus was that it was a risk worth taking.

It was at that meeting that I first met Mike Rossiter, the BBC producer who would actually be making the film. He had flown in from London for the meeting.

Mike already had a proven track record when he was asked to participate in our project. At the time he had been with the BBC for only a few years, but he had many years of experience as an independent producer and had produced a number of documentaries for the Learning Channel, Nova, and the BBC.

Mike's accent is working-class London, and his working style is absolutely no-nonsense. At first glance, his demeanor seemed testy, bordering on arrogant, until I got to know him a little better and realized that was just his way. He is all business, no bull. He was not going to choreograph a successful expedition; he was going to document whatever happened. If I fell on my face, that would be the show.

Mike knows what he wants and makes it happen. The BBC is very demanding as far as the planning that it needs to see on paper is concerned. We prepared risk-assessment reports, documented safety procedures, and outlined for Mike exactly how we would be going about our job.

From our discussions, we were able to determine that an expedition of a couple of weeks would be enough for us to finish our work and for Mike to get the footage he needed to make a documentary about the wrecks. Besides, it was all the money that was available.

Unfortunately, as we were working out our plans, Charles Brewer was making his own plans. His plans apparently included spending significant amounts of other people's money. He e-mailed Mike Rossiter and me to outline what he thought was necessary to secure the permits. Mike was not happy when he read it. At the heart of the problem was the fact that Charles was angling to get the BBC to sponsor the expedition under his direction—and on his financial terms. Mike e-mailed me that Charles's demands were "causing alarm and despondency among my executive producers."

For someone who had mysteriously appeared on the first trip, Charles had come a long way. He had managed to get in a position where his interference was threatening the production, which would have been disastrous for the project. At Mike's urging, I contacted Charles and explained the situation, telling him what the production company would—and would not—do. He was not pleased to have his plans thwarted, and I imagine he saw all manner of plots forming against him, but, in the end, he relented. After all, he was not going to get on television otherwise.

It was not a good start. And it cemented an enmity between Charles and Mike that would hang over the entire expedition.

Mike has enough experience working in foreign countries to know where the pitfalls lie. Though Charles had managed to usurp the permitting process—to make himself indispensable—Mike understandably wanted his own person on the job as well, to insure that all things went smoothly. He hired a Venezuelan named Antonio Casado to help us out with arrangements in Caracas. Antonio is a television producer in his own right, having worked for years with Venezuelan TV. That experience was crucial, because it meant that Antonio knew what we would need for television production and understood the specific problems we might encounter. People like Antonio are known as "fixers." Antonio went to work right away, and he was a godsend.

Antonio went to the Ministry of Defense and the various other ministries to see what we would need. After much legwork, he called Mike and told him that, in his view, all the BBC needed was filming permits. Then he added, "Of course, you should also have some expedition permits."

"What are they?" Mike asked. Expedition permits seemed to be in Charles Brewer's domain, and he had told us he was handling them. I was very concerned, given our competitors' insistence that they alone had the rights to explore Las Aves. But Charles insisted that there was no problem, that the permits were in order.

Charles was well connected; his brother knew this person, his mother knew that person, he talked to the right government officials. Charles sent us copies of permits. Pedro Mezquita, who works in a major law firm in Caracas, said everything looked fine.

Charles and Antonio were in contact with one another, coordinating their efforts, and Mike assumed that they were handling affairs in Caracas. Charles sent an e-mail stating that he "had obtained the permits from the Minister of Defense," and the filming permits were finalized three days before we were due to fly out to Venezuela. From his home in London, Mike called Charles and asked if any other paperwork might be required. "No, no, no. Don't worry about it, we'll sort it all out," Charles told him.

We had spent the summer working the wreck of the *Whydah*. It was a great season for us. Not only did we bring up an impressive collection of artifacts, but we also located a sixty-foot section of hull, the only wooden section thus far. The National Geographic Society was there to capture the event for both television and their magazine.

The working season is short off Cape Cod, however, and by early fall it was over for the year. On October 21, 1998, we were ready to go south.

I woke up at seven in the morning, roused by what sounded like a grizzly bear trying to knock my building down. It took me a moment to realize it was Stretch Gray, the six foot ten, three-hundred-pound captain of the *Vast Explorer*, pounding on the side of my home at the end of Macmillan Wharf. He was our driver to the airport. I climbed out of bed.

We made it to the airport forty-five minutes before our scheduled departure. We had a small mountain of gear and personal luggage—fourteen bags of it.

By some miracle, our gear and our team all made it onto the airplane. Before we left, I gave everyone a quick primer on what to expect in South America. I have spent a lot of time in the Third World, but Caracas has a particularly sharp edge to it. I was determined that none of my team would get cut. We boarded the plane. Next stop, Venezuela.

13

A Pleasant Accident

We were certainly not the first to come poking around the wrecks at Las Aves. Someone else, perhaps the Florida treasure hunters, had been there since our first trip to the island. Near one of the larger wreck sites we could see where something had been salvaged from the reef. Anything that has sat on a reef for three hundred years becomes embedded in the coral, and when it is wrenched out, it leaves a big, fresh hole.

At this site someone had done just that. From the size and shape of the hole and the artifacts around it, it looked as if it was a cannon that had been taken, perhaps one of the valuable bronze cannons. I was sorry to see the site vandalized in that way. And looking at the damage to the reef caused by the removal of one gun, I began to wonder if any excavation would be worth the environmental cost.

Neither of us was first on those wrecks. The first salvagers showed up while they were still fresh. One of them was Captain Thomas Paine.

Not to be confused with the Tom Paine of *Common Sense* fame, Thomas Paine the pirate holds a special place for me. Paine is one of the living links between the early era of the golden age of piracy, the middle to later seventeenth century, and the later period that ended around 1720. By the beginning of the eighteenth century, however, he was a respectable citizen. He was also a pirate kingpin who had numerous criminal connections among those pirates and smugglers who made my native Cape Cod home.

THE FIRST SALVAGER

Captain Thomas Paine was in some ways the polar opposite of the Chevalier de Grammont. Rather than starting life in the upper strata of society, Paine began as a simple seafarer, a most common profession, particularly in New England, where he grew up.

Thomas Paine is believed to have been the same Thomas Paine who was born on Martha's Vineyard in 1632. His father died young and his stepfather was Thomas Mayhew, the famous governor of that island. He became a sailor and left the island no earlier than 1647. The exact circumstance of his having turned pirate are unknown, but he is believed to have served with L'Ollonais or Morgan, or possibly with both.

While we do not know how he first arrived in the West Indies, it was definitely not as a commissioned officer in anyone's navy. I know what Martha's Vineyard was like in the old days and for Paine it took determination and hard work, not to mention a good deal of luck and money, to rise above his humble beginnings. Only after a long and successful career as a privateer and pirate would he become a pillar of society.

Paine was already established as a filibuster captain by the spring of 1678. Since the proposed attack on Curaçao involved so many of the buccaneers, it is probable that Paine was a part of it, and like the others found himself on Las Aves. Paine's ship, which sported six guns, would have had a shallow enough draft to avoid wrecking on the reefs as the larger ships did.

From Las Aves, Paine appears to have accompanied de Grammont and the rest to Maracaibo in June 1678. Much of the documentation is found in French colonial records, and the vagaries of seventeenth-century French make it impossible to know for certain if such spellings as "Gouain" refer to Paine. If Paine did go to Maracaibo, he stayed only for the initial festivities. In the summer of 1678, Paine turned up at Las Aves to refit and careen his ship.

Careening was a laborious process by which a ship was emptied of guns and stores, run up on the beach on a falling tide, and hove down nearly on her side. As the water dropped away the vessel was left high and dry and the crew was able to scrape weeds and barnacles from the exposed side. The ship was refloated on an incoming tide and hove down on the other side to complete the cleaning. The whole process was labor-intensive and left the vessel defenseless, but it was necessary to keep the hull clean to maintain the kind of speed needed by fast-moving pirates.

Not only did Las Aves have beaches on which to careen, it was the perfect place to refit and replace old and worn gear. With the French fleet's disaster only a few months old, Las Aves must have been like a free open-air chandlery, with all manner of spars, rigging, timber, and stores handily scattered over the beach.

Careening a pirate ship

Paine sailed into harbor and then prepared to heave his ship down. The first step was to unrig the ship: to take down all of the masts and yards, sails and rigging so that they would not hamper the heaving down or be damaged in the process.

So Paine's ship was completely immobile when, by sheer bad luck, a Dutch ship of twenty guns hove into view. The Dutch, of course, were as aware as the pirates of the booty to be found on Las Aves, and this ship had been dispatched from Curaçao to retrieve it.

The Dutch knew exactly what this strange ship was about. They sailed to within a mile of her and opened fire, to little effect. At last they ceased fire and anchored, most likely because night was falling and the Dutch captain did not care to get tangled up in the reefs after dark.

Instead, the Dutch made preparations to move the ship in closer the next day by means of warping, a process that involved dropping an anchor out ahead and hauling the ship up to that anchor. There was time enough. The Dutch captain knew that this interloper, with his ship stripped of masts and sails, was not going anywhere.

It seemed hopeless, and, indeed, it probably was, but Paine was a man who would not go down without a fight. He began to set up as good a defensive situation as he could although he knew it would only delay the inevitable. He and his men were outnumbered and out-gunned.

Then, ironically, salvation came in the form of another Dutch vessel, a sloop that sailed in near the west end of the island and unwisely dropped anchor. Once night had fallen, Paine and his men went out in two canoes and boarded the sloop, taking her before an alarm could be raised. Then, tiptoeing out to sea, they abandoned their old and tired ship to the Dutch man-of-war. Thus bad luck had become what Dampier described as "a pleasant accident."[1]

Having made their narrow escape, Paine and his associates continued their cruise against the Spanish. He and his men joined forces with two other vessels. One was commanded by an English buccaneer named Wright who sailed with a French privateer's commission. The other commander was a Dutch captain, Jan Willems, better known as Yankey, apparently a mangled version of the Dutch "Jan."

Little is known about Yankey's prior life, but in the years to come he would make his presence felt on the Spanish Main. Together he, Wright, and Paine prowled the seas for Spanish prizes.

THE SACK OF RÍO DE LA HACHA

In 1680, Paine and Willems were among those who staged a raid on the city of Río de la Hacha, about one hundred miles east of Cartagena in present-day Colombia. Along with plundering the town in high style, the pirates kidnapped for ransom a number of important citizens, including the governor of the province, Vincent Sebastian. They also hit the nearby town of Santa Marta. With their prisoners aboard, the freebooters sailed for Jamaica, and from there issued their demands.

News of Paine's outrageous behavior reached the very highest levels, providing a perfect example of the official wink and a nod occasionally directed toward piracy. Paine's actions were legitimized by a French commission, thus making Paine's attacks technically French aggression. But there were English fingerprints all over this action despite the fact that Spain and England were at peace.

Not only were Paine and Wright English, but they transported Governor Sebastian and the others to the English island of Jamaica and held them there. Paine demanded four thousand pieces of eight for the return of the governor as well as the release of an unnamed French pirate held prisoner at Cartagena.

To make matters worse, the demand for the release of the pirate and the money arrived in Cartagena aboard a barque from Jamaica. And the final straw, the demand itself was cowritten by the governors of Jamaica and Tortuga. The outraged Spanish ambassador in England, Don Pedro de Ronquillos, wrote to Charles II:

> For though it be said that Frenchmen did it [invaded Santa Marta], yet it is certain that English were with them, and that they sailed with their prisoners to the port of Jamaica, where the governor ought to have chastised your Majesty's subjects and not consented to demand a ransom for them.[2]

All that year, Paine and Wright's activities continued to plague the Spanish. Again Don Pedro wrote to Charles II:

> [T]he captain of the Armado de Barlovento . . . [experienced] the infraction of the peace, in that a small vessel under his charge was taken by [from] him in company of an English frigate, a bark and a flat-bottomed boat. This is affirmed in the declaration of the

A barque

inhabitant of Margarita aforesaid, who says that the captain of one ship was called Thomas Pem [Paine] and of the other Heohapireray [possibly a corruption of Wright], both English, and that the men were also English, with a commission from the French Governor of Tortue.[3]

The commission alluded to is presumably the one issued by Governor M. de Pouançay to all of the buccaneers hired at Tortuga for the attack on Curaçao. If this is so, then it is unlikely that a two-year-old commission would still be considered valid by any honest jurisdiction. It was, however, good enough for the buccaneers. Since piracy of the late seventeenth century had a decidedly political flavor to it, it was probably good enough for English and French authorities, too, so long as grand theft was confined to the Spanish.

Don Pedro goes on to describe the wild spree on which Paine and his cohorts were engaged:

> These same and other pirates also landed in Honduras, and after many insolencies plundered the King's magazine and, among other things, carried off a thousand chests of indigo [a valuable purple dye] which they are known to have sold in Jamaica as they do the rest of their booty and prizes. These are not the only insolencies of these pirates; they infest the Isles of Barlovento, and have plundered Porto Bello, the most important city on the coast.[4]

RENEWED ALLIANCES

While Paine and Wright were rampaging across the Spanish Main, the Chevalier de Grammont tarried in Tortuga, enjoying the fruits of his piracy. In May 1680, a year and a half after Maracaibo, he was once more ready for action.

The Chevalier met Paine and Wright at Isla La Blanquilla, about two hundred miles west of Grenada, part of present-day Venezuela. This meeting was fortuitous, at least for the buccaneers. Paine and Wright decided to join de Grammont on his latest venture, perhaps the boldest and most audacious he would ever undertake.

The target was La Guaira, the port of Caracas, a place well protected by two forts and with cannons mounted on the city's walls. Dampier reports that the Frenchman was acting on the strength of an old commission granted him by de Pouançay. De Grammont, like Paine, was using his old papers to give his activities the thinnest veneer of legitimacy. For the buccaneers, it was good enough.

14

The Sack of Caracas

Oh the palms grew high in Avès, and fruits that shone like gold,
And the colibris and parrots they were gorgeous to behold;
And the Negro maids to Avès from bondage fast did flee,
To welcome gallant sailors, a-sweeping in from sea.
— "THE LAST BUCCANEER"
Charles Kingsley

JUNE 26, 1680
LA GUAIRA, VENEZUELA

On the night of June 26, 1680, the buccaneers came ashore at La Guaira with a mere forty-seven men. The pirates were vastly outnumbered. Rather than attacking, they slipped unseen into the city. They infiltrated the garrison and managed to capture the 150 soldiers stationed there without raising an alarm. That morning the civilians woke as usual, only to discover that their city was occupied by buccaneers.

De Grammont and Paine—bold but not stupid—knew that their position was weak. They began to loot as fast as they could, knowing that more Spanish troops would soon be on the way. As successful as the buccaneers traditionally were in such raids, they knew that there was only so much that forty-seven men could do.

Nor were the 150 soldiers they captured the only troops in the area. A Spanish officer, Captain Juan de Laya Mujica, and his company managed to escape. The captain sent a warning to Caracas of the buccaneer attack, and at the same time rallied what soldiers and militia he could from the area around La Guaira.

When word of the raid reached Caracas, the usual panic ensued. The inhabitants loaded valuables on wagons and sent them inland. At the same time, the governor, Francisco de Alberró, organized a large contingent of militia and marched them off toward the occupied port city.

By daylight, it became clear to the Spanish that they had been attacked by a very small band of filibusters. Captain Juan de Laya Mujica, apparently an active and responsible officer, was emboldened by this. He led his troops in a counterattack against the pirates.

In the face of this assault, knowing that Governor Alberró's forces, now numbering two thousand men, were marching for La Guaira, de Grammont and Paine decided it was time to go. Still, the Spaniards did not rout the pirates despite their fifty-to-one advantage in the area. De Grammont and the others fought an organized rearguard action, retreating to the harbor with booty and prisoners and then out to their waiting ships in an orderly fashion.

De Grammont (and most likely Paine) personally covered the retreat of the men, holding off the Spanish troops. Dampier reports, "This movement was executed with difficulty, and for two hours de Grammont with a handful of his bravest companions covered the embarkation from the assaults of the Spaniards."[1]

Of the forty-seven men who attacked La Guaira, only eight or nine were lost, but de Grammont was nearly one of them. A lucky Spanish swordthrust severely wounded him in the neck, and the Chevalier barely escaped with his life. Despite the kidnapping of the governor of La Guaira and many other prisoners, the pirates' take was small, particularly considering the hazard involved. In less than twenty-four hours they had probably endured more desperate fighting than they had in six months at Maracaibo.

De Grammont and the rest proceeded to Las Aves, where the Frenchman intended to convalesce. In the meantime the Chevalier turned command of the small buccaneer squadron over to Paine.

While at Las Aves, de Grammont reported that on "the 2nd of August I left the command of the King's subjects to Capt. Pain [*sic*] with orders to conduct them to the coast [of French Hispaniola] and give an account of our actions to the governor. . . ."[2] Just which monarch de Grammont thought them to be subjects of is not clear. Presumably the Sun King, Louis XIV, from whom their commissions originated. Whether or not Louis was happy to have them as subjects is unknown.

AN ILL-FATED EXPEDITION

A year or so after the La Guaira raid we find Paine ranging the Spanish Main. In May, 1681, a number of captains came together at Springer's Key in the Samballoes Isles, near the coast of Panama, eager to cooperate in a joint raid. This was typical of the ad hoc nature of filibuster armadas. Present were ships and men of English, French, and Dutch extraction, most of whom had been "on the account" for years and many of whom had been among those flung up on the beach at Las Aves three years earlier.

Of the English and Dutch captains, there was Paine, with a ship of ten guns and carrying one hundred men; Captain John Coxon, similarly outfitted; and Paine's old consorts Wright and Yankey Willems.

Over the next decade, Yankey would become one of the foremost of the buccaneers and participate in nearly every major filibuster engagement in the region. For this expedition Yankey commanded what was essentially a very large sailing boat, called a *barcolongo,* carrying four small cannons and manned by a crew of sixty men, including Englishmen, Frenchmen, and fellow Dutchmen.

The French captains included a Captain Archemboe and Captains Tucker and Jean Rose. Little is known about them. Their crews numbered around 150 men, about half the size of the English force.

One of the political realities illustrated by this meeting was the fluid nature of English and French alliances. Despite on-again, off-again warfare for more than a century, the filibusters of the two nations could work together and even accept commissions from the others' government. What unified them was a universal hatred of the Spanish, and particularly of Spanish attempts to retain an iron grip on the riches of the Caribbean.

To this gathering of filibusters came another French captain named Tristian. Although his ship was undermanned and in poor condition, he had on board William Dampier, the great adventurer, filibuster, and author, fresh from the South Seas. Tristian had recently rescued Dampier and his shipmates from the nearby La Sound's Key after they

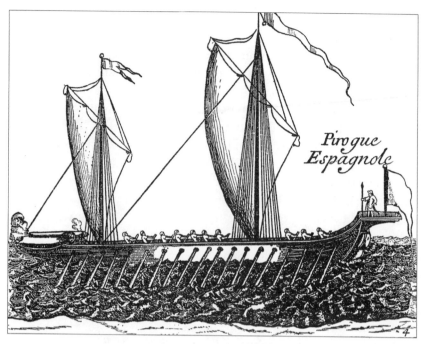

A *barcolongo*

had abandoned their former captain in the Pacific and marched back over the Isthmus of Panama.

Dampier stayed with the buccaneers, and we are fortunate that he did and that he lived to write about it. Dampier provides us with a wonderful firsthand account of the organization of a buccaneer raid.

Paine's consort, Captain Wright, had been sent to the Panamanian coast to find a prisoner from whom they might gather intelligence. He returned with two prisoners and their canoe laden with flour. The captains assembled aboard Wright's ship and interrogated the prisoners as to the condition and strength of the city of Panama. Their plan was to march overland to the town, using the wild and often hostile San Blas Indians as guides.

The captains took under advisement the intelligence gathered from the prisoners and fell to discussing where they might mount an attack, whether Panama or elsewhere. Here again is classic pirate democracy in action. For seven or eight days they discussed their plans, meeting every day to try to find a mutually agreeable course of action.

The men at Springer's Key were very knowledgeable about the Spanish Main. As Dampier expressed it, "The privateers have an account of most towns within 20 leagues of the sea, on all the coasts from Trinidad Island down to La Vera Cruz: and are able to give a near guess to the strength of them. . . ."[3]

It was decided at last to mount a raid on a town lying on Carpenter's River. The river allowed the filibusters to attack the town in boats, thus avoiding the horrible and often fatal march though the Central American jungle required to reach Panama.

The pirates weighed anchor and set sail for the small, uninhabited island of San Andreas. They intended to fashion dugout canoes from the abundant cedar there for their attack on the river town, a most ambitious plan.

The English ships already assembled at Springer's Key were overmanned. Paine's ship of a mere ten guns could not have been very big, certainly not to ship one hundred men aboard. Dampier was forced to sail aboard the French captain Archemboe's vessel of eight guns, which was undermanned with just forty men.

Though they might be able to put aside political differences on the grand scale, the English and French were not so quick to set aside hard feelings on the personal level. The Englishman Dampier was none too happy about shipping with Frenchmen, nor was he overly impressed by the quality of seamanship aboard.

The first day out, the fleet kept company, but that night a hard gale blew from the northeast, and by the following night, the ships had lost contact with one another. Dampier was disgusted by his shipmates' feeble efforts to combat the storm. "Indeed we found no reason to dislike the captain; but his French seamen were the saddest creatures that ever I was among; for tho' we had bad weather that required many hands aloft, yet the biggest part of them never stirred out of their hammocks, but to eat or ease themselves."[4]

Despite the crew's relaxed, casual attitude toward seamanship, the vessel survived the storm and by the fourth day reached San Andreas. Of the ten vessels that had started out, only Wright's was there when they arrived, but Wright had managed to take a prize. It was a Spanish tartan, a small armed vessel the buccaneers had fought for an hour before taking. The tartan, as it turned out, belonged to a fleet of small men-of-war sortied specifically to rout out the pirates assembled in the Samballoes Isles, of whom word had reached Spanish authorities.

Dampier and his fellow South Sea adventurers saw the new prize as

The tartan

an opportunity for them: "We that came over land out of the South Seas, being weary of living among the French, desired Captain Wright to fit up his prize the tartan and make a man-of-war of her for us. . . ."[5] Wright, who did not share Dampier's dislike of the French, at first refused, and only relented when told that the South Sea men would build canoes and leave their company before they would sail with Archemboe again.

The men remained at San Andreas for ten days, waiting for the others to arrive, but only Captain Tucker turned up. At last, they left the island, concluding that the others must have been blown too far to leeward to claw their way back.

Searching for their companions, they eventually rendezvoused with Yankey Willems. Willems had met a fleet of Spanish armadillas, small men-of-war. The tartan that Dampier and the others insisted on having had been a part of this fleet. In the ensuing fight, Willems and his consorts had been scattered. And so, with their carefully laid and much debated plans now in ruin, the various filibuster captains went their separate ways, sailing off in pursuit of other ventures.

Captain Thomas Paine remained in the area of Boca del Toro in the northwest corner of Panama, where he had been blown by the gale that had scattered the fleet. Deciding that his ship could not go on without attention to the hull, he emptied the vessel and careened her on a convenient beach.

Unfortunately for Paine, he chose an area where the Indians had never made peace with the white men, nor had they been beaten down by force. Dampier explained, "The Indians here have no commerce with the Spaniards; but are very barbarous and will not be dealt with."[6] One night, while some of Paine's men slept in a tent ashore with their weapons at their sides, a band of Indians quietly decapitated three or four of them and slipped back into the jungle. All in all, it had not been a good voyage for Paine.

With the storms and the armadillas and the leaking hulls and the deadly natives, it is little wonder that by October 1682, Thomas Paine was tired of piracy. He decided that a more prudent course might be to turn pirate hunter, and for that work he applied to Sir Thomas Lynch, the royal governor of Jamaica, through an intermediary named Clarke.

In November 1682, Lynch wrote to the secretary of the Council of Trade and Plantations Sir Leoline Jenkins, informing the secretary that "one Captain Clarke, a very honest useful man, solicited me about one Payn [sic] in a barque with eighty men. He told me Payn had never done the least harm to any, and if I would allow him to come in, he would engage to bring in or destroy these pirates."[7]

Lynch seems to contradict himself when he claims that the honest Captain Clarke assured him that Paine "had never done the least harm to any" yet acknowledges that he needs the governor's permission to "come in." Perhaps "the least harm to any" in this sense meant any Englishmen or their allies.

Lynch further reveals his own understanding of Paine's former trade when he goes on to say, "I thought this likely and advantageous from creating division among the pirates so I accepted the offer and hope *per fas aut nefas*[8] to put down these destructive rogues." Clearly Lynch believed in the old saying "Set a thief to catch a thief." He knew full well that Paine was a pirate and that other pirates might be intimidated by having one of their own coming after them.

In the fall of 1682, after at least three years of piracy, Thomas Paine accepted the governor's commission and set about the job of pirate hunter, intending to live an honest life and forswear piracy forever.

Or so it seemed. . . .

15

Ready for the Ends of the Earth

Caracas is one of the most dangerous cities in the world. It was once a verdant jungle in a thousand shades of green—until the conquistadors came. Now, five hundred years later, stripped of trees, wildlife, and native vegetation, the landscape has the appearance of well-baked piecrust.

Venezuela is the largest oil-producing country in the Americas and a member of OPEC. It has a New York City skyline—a big money skyline. Beneath it, however, is a scene reminiscent of Calcutta. We were told that there were, on average, a dozen or more murders a week.

The contrast between unimaginable wealth and unimaginable poverty was evident on the drive from the airport to the city. Cardboard houses lined one of the finest roads on which I have ever driven. It is a road built for expensive European cars and military vehicles, a road built with oil money for the rich and powerful.

It is a myth that there is a middle class in Caracas. Either you have money or you are poor, and the vast majority of people in Caracas are desperately poor—and without hope.

I am reminded of a story a Venezuelan fisherman told me about a crab trap teeming with crabs at the bottom of the Venezuelan sea.

One crab manages to escape and sits for a moment on top of the trap. From the tangle of crabs below, his friends call up to him, "Save us, Brother, open the latch!" But the crab does nothing. Again, the trapped crabs scream from below. "Help us, Brother, or we will surely boil in the cook's pot." Finally, the free crab just scuttles away. "We are doomed; why didn't he open the latch?" asks one of the crabs of another. "There is no reason—this is Venezuela, my friend."

To have *esperanza,* the Spanish word for "hope," is now an anachronism in Caracas.

When we arrived, the country was preparing for elections, though not the typical elections where political power is simply circulated within the ruling class at suitable intervals. A new politician, Hugo Chávez, was running for president, and he was different. Chávez is from the barrio, a man who worked his way up and out of poverty. Unlike the old aristocracy, he is dark-skinned. He is a man who said he was serious about changing the order of things, and both his friends and his foes believed him.

It looked as if Chávez would actually be elected. The aristocracy was naturally terrified at the prospect, and the country was tense. Perhaps it was just our American perspective, but chaos seemed to rule the day. We saw a girl run down by a motorcycle, and the driver continued on as if nothing had happened. Police officers seemed to outfit themselves according to their own taste, and the only thing they had in common was that they all dressed like army generals.

Caracas would not have been much of a problem, except that my crew still needed to buy last-minute supplies. I wasn't happy to have them wandering around, but there was no choice. We needed batteries and other gear.

Our only concern was to get that gear together and get out of Caracas as soon as possible. The pending election was not our concern.

We were in Caracas a long five days waiting for the different parts of the project to gel. Charles Brewer and Antonio Casado had assured us that our permits were in order, but the BBC needed more assurance, which meant pounding on a few more bureaucrats' doors. It was also our last chance to double-check that we had everything and that everything worked before we left civilization. Las Aves might not have been the end of the earth, but you could see the edge from there.

On our first night in Venezuela, Charles invited the entire team to his house for a grand bash, a buffet dinner and a big party to launch our expedition. "Don't bother to eat," he told us. He had a big spread planned. We were all more than a little curious to see where Charles Brewer lived.

We drove through the city and then started climbing up into the hills where the wealthier people had homes. We came to a gate through which we had to pass: a dividing line between the haves and the have-nots. We went through and continued on up the hill.

At first, we figured we were heading to a grand home in an affluent subdivision, like the gated communities in America. To our surprise, the driver continued on up, higher and higher, up the mountainside. As the road grew steeper, it turned into a series of switchbacks, zigzagging up the face of the mountain.

The houses became fewer the higher we climbed until, finally, there were no other houses, only the road and the black night. It was surreal. We were all looking out our windows with growing concern. Off the road, it was a straight drop down, and, of course, there were no guardrails.

Charles Brewer's house was at the very top of the mountain. To the east lay the city of Caracas, a great valley of light spread out like phosphorescence in the ocean at night, then cut off in a sharp line where the city met the sea. To the southwest, there was only blackness. The jungle stretched away virtually unbroken for hundreds of miles toward the Guarico, a tributary of the Orinoco River. It was magnificent.

The house itself was worthy of its extraordinary location. It was framed entirely with massive posts of an exotic hardwood such as lignum vitae. The walls were stucco and stone, constructed with the greatest care and craftsmanship. The house looked like a grand hunting lodge in Kenya, something built for wealthy British imperialists of the nineteenth century. Wrapped around the front of the house was a wide veranda, complete with hammocks and wicker furniture, where gentlemen could smoke cigars and socialize. It was everything I would have expected of Charles.

We were very late when we arrived. The path from the car park to the front door was winding and overgrown. The air in Caracas is terribly polluted. But up there, the breeze bore the scent of the jungle, the breath of flowers and new life, mingled with the odor of ancient trees, wood smoke, and decay.

Charles met us at the door. He was tan and fit and dressed neatly in pressed khakis. He proceeded to show us around the house. Calm and personable; the perfect host.

He had been waiting patiently for us to arrive, and greeted us with the announcement that "Terminator," his pet killer ant, had escaped its cage. "Be very careful," he said, "his bite, above all others, is feared by the Yanomami." This rare species of ant is a nasty character, a solitary creature an inch or so long whose bite, Charles said, is feared by the Amazonian Indians more than the bite of a bushmaster. In fact, Charles told us how he had once attempted to bash his own brains out against the floor after having been bit, and that he had heard stories of Indians roasting a bitten arm or leg to ease the pain.

"Ah, you can be sure he is watching us at this very moment." Then, with the composure of a great white hunter, Charles took control of the situation. "Follow me," he said, as he led his anxious dinner guests about the house in search of the little killer. We passed deadly snakes in cages, spiders in shadow boxes, case after case of mounted butterflies, and the skeletal remains of countless creatures—a macabre museum of natural history, all placed with the deliberation and care of a Hollywood set designer.

This was Charles at his finest. In control of a group of potential investors and media people whom he was directing through a maze of psychological traps, which he had baited and set. "Ah, there you are." Indeed, pinned to the back of a shadow box was the horror of the Amazon rain forest. Charles was the first to laugh. "It is a great joke, is it not?"

We retired to the sitting room, where there happened to be a boa constrictor hiding under our couch. "Come here, Imperator," he called, giving it its scientific name, as he picked the reptile up and looped it around his neck.

I was thoroughly amused. Though, like the sales pitch of a used-car salesman that had worn thin, Charles's shtick had lost its spontaneity. Indeed, this was not the first time "Terminator" had come back to life.

But what I saw next startled me more than if Terminator had been sitting on my lap. There, on the mantel, as conspicuous as John Wayne in a tutu, was the pot that had been found on the first expedition. The one that Charles had said he would turn over to the Minister of Culture.

Perhaps he noticed me staring at it. He quickly asked Margot if we'd like to see the grounds. We followed him beyond the lawn down a path through low, scrubby bushes. Charles explained he had been having trouble with wild dogs killing his pet deer. We arrived at the spot where he had caught and killed one of the dogs the day before.

Going into careful detail, as if we were grad students and he were the instructor, he described how he went about catching the wild dog.

"You must first determine the paths he uses. Then take a big strong fishhook, like the kind used to catch tuna. Bait it carefully to hide the hook, but not too much that the hook won't set quickly. Then suspend the hook from a tree, high enough that when the dog snatches it, he will hook himself and dangle off the ground." The thought of a dog with a hook through its mouth, jerking in anguish, was intended to let me know that Charles was a cunning and careful trapper. That he knew tricks I had never dreamed of . . . I got the message.

If I still had any illusions about going into the jungle with Charles, he had just removed them. I needed to get back to my element, the sea. The sea can work against you, but it can work with you as well. The Navy SEALs know this. While the sea is powerful enough to tear ships in half, it can also afford protection, if you know its ways.

When we returned to the big, open living room we found that the "banquet" consisted of cheese, crackers, and local wine. "This is one hell of a banquet," Mike Rossiter muttered to himself.

"Bloody hell," I said to Pedro Mezquita with my best British accent, "this is going to be a long voyage!"

The "buffet dinner" didn't bother me. I just ate a lot of cheese, but Mike was really upset. I could see why it would bother him—it was his job to look after his crew—but I thought he was overreacting. He was spitting mad. Worse, his camerawoman had fallen off the porch and twisted her ankle and might be out of the game. Mike blamed Charles for the porch being unsafe. Mike was predisposed to finding fault with Charles after all the problems Charles had given him about preparations for the expedition. Charles made it easy.

Charles was never more than two steps away from a loaded gun. There were guns propped up in every corner, like the scene in the blockhouse in *Treasure Island*. With someone like Charles, it is hard to know how much is sensible precaution, how much is paranoia, and how much is showmanship. He referred to the trackless jungle beyond his house as his "backyard" and his "escape route." When "they" came for him, as he put it, he intended to disappear into the jungle.

Indeed: "Just because you're paranoid doesn't mean they're not out to get you."

This was Caracas. Charles had a wife and children to protect. Paranoid or not, I too would have armed myself.

It was after the party got rolling that Charles took me aside. "Barry, I must talk to you. We have a slight problem."

"Really? We have the permit, right, Charles?"

"I am not sure we have the permit. Don't tell Mike."

To say I was stunned is an understatement; I did *not* want to hear this.

"Charles, we have the BBC and the Discovery Channel here. I have spent a fortune. I have my team with me, and we've hauled all this gear down here."

"Don't worry, everything will be fine. We'll get the permits tomorrow, or the next day," he assured me—which I did not find reassuring. In all the phone calls with Mike and me, all the e-mails he'd sent, he had minimized any possibility that all was not in order.

To be fair, Charles did have *some* permits. In fact, before we headed out to Las Aves, Mike Rossiter and his crew videotaped Charles going from one government office to the next, talking with the officials there and getting permits. Some of those meetings even ended up in the documentary. But the officials with whom Charles talked might not have been the ones whose permission was most needed. In a country like Venezuela, we could not be sure who had the ultimate authority. Was it the admiral, or the general? Or the general's nephew?

Pedro Mezquita told me that the Minister of Defense would not issue a permit because he felt "left out." But after Pedro invited him out to the site for a visit, he felt much better about our project and agreed to issue the permit.

Another part of the problem might have been the competing group that also had permits, issued by the navy, to work Las Aves. There was circumstantial evidence to suggest that whoever had issued the navy permits did not want to see us with permits as well. It was like competing travel agencies had sold two people a ticket for the same seat on a flight.

This was an area where Mike's experience paid off. Before going to film in a foreign country, Mike makes a point of establishing a relationship with an official from that country, generally its ambassador in London. He considers it part of the courtesy one should extend to a host country.

It is also a safety issue. As he put it, "We were going to be on a boat on an atoll quite a long way from anywhere, with limited resources, particularly limited medical resources. If you suddenly want to call in an emergency and get a helicopter from the army or navy, I always think it is helpful if you have already made contact with that government's representatives."

He was right, and that forethought would prove useful. The Venezuelan ambassador took a personal interest in the project. Though Mike went to the embassy expecting only to speak with one of his assistants, the ambassador himself called Mike into his office to discuss our plans. He was very interested and enthusiastic—a good man to have on our side. We had no idea then just how important that contact would be.

16

Tools of the Trade

In Caracas we acquired last-minute supplies, organized our research and charts, and set up and tested our gear. For this we used the cafeteria in the hotel, which we called the Blue Room, for obvious reasons. We were lucky in our accommodations. If Caracas itself is generally unsavory, the hotel was lovely and the staff were very friendly and accommodating.

Cathrine gave a demonstration of underwater mapping and drawing techniques to the assembled team. We had worked together before, but I felt it was a good idea to make sure we were all on the same page before heading out into the field. We wanted to be as organized as possible so we could make the best use of our limited time.

Some of the communications gear was new, and we had to make sure to test it before we had to use it on the site. We went over the scuba gear and other equipment as well, to check that it was all in order. There would be no sending out for parts at Las Aves.

The equipment we would use fell into three basic categories: dive gear, communications gear, and search and mapping equipment.

Dacor, a company I have been using for thirty years, donated a good deal of gear to the expedition, wet suits and regulators and such, though some of the team preferred to wear their old, broken-in wet suits. In

Max Kennedy searching for clues

truth, we would have preferred to wear no wet suits at all, the water was so warm. That was not an option, not with the sharp coral of the reef.

Communications are very helpful in underwater archaeology, and in my opinion they are one of the best safety tools you can have. If the person in the water can't communicate, he can't tell you he is in trouble. I have had several situations in which communications saved the life of the diver. Once, working in the East River in New York City, our diver got himself wrapped up in a shipwreck. His umbilical, which attached him to the ship, became so tangled that he could not get it undone, and he could not get out of the secure harness by which he was attached to the line. If he had not been able to call us, he might not have gotten out before the tide switched directions. As it was, we were able to send another diver down, who followed the umbilical to the trapped man and got him free.

The closest call we had was diving on the *Whydah*, when Chris Macort came within inches of getting killed. We had the *Vast Explorer* anchored near the shoreline. My smaller boat, a Boston Whaler named the *Andrew Crumpstey*, was anchored by the bow and tied by the stern to the *Vast*'s windlass. I was in the pilothouse, working on the computer. Chris was in the water, working at the bottom of a pit we had excavated, and the rest of the crew were at the stern, monitoring his progress.

Suddenly, over the com, Chris began screaming as if he'd been scalded with boiling oil: "Help me! Help me! Something's got me! Help, before I'm killed!"

Cathrine is normally unflappable, but she came into the pilothouse in a sheer panic and screamed at me that Chris was trapped, that he was being killed.

They were due to leave for Scotland in a few days to be married.

A hundred thoughts rushed through my mind in less than a moment; the first being that the big white shark that had been rumored to have been killing seals in the vicinity had Chris. I ran below to get my gear on. Then, seeing the *Crumpstey* off the bow, I realized the anchor line on the *Vast* had parted. Held only by the smaller anchor of the *Crumpstey,* we were slowly drifting into the breakers less than fifty feet away.

The captain, a man who had worked with me many years before and was filling in for Stretch, was frozen in place, useless, with a look on his face as if someone had just stunned him with a rubber mallet. I tried to pull Chris in on the safety line, but he had been pulled out of the pit and swept into the current, and was now on the downcurrent side of the vessel.

I did not want to risk pulling him all the way under the boat, as the water was getting shallower every moment, threatening to crush Chris when she bottomed out. I tried to calm Chris over the com. I told him to get out of his harness and swim free. He could not do that. His hands were too cold to manipulate the nylon straps through the D rings. We were a couple of minutes from going ass-over-teakettle into the breakers . . . there could be no mistakes.

In the early seventies, I worked off Martha's Vineyard with the well-known swordfisherman Greg Mayhew. Greg had learned from his father to never go to sea without carrying at least one very sharp knife on his person at all times. Greg's brother, Skip, had been fouled in a line of lobster pots once and was nearly taken to the bottom. It was a lesson he never forgot.

I pulled Chris to the surface by his safety line, grabbed my Benchmade automatic knife from its sheath, and cut through the big straps of the harness as easily as if they were wet noodles. Communications, a sharp knife, and quick reactions saved Chris that day.

Safety aside, communications is also important to the work we do. The divers in the water have to be able to speak to one another to coordinate their work, and to the people in the boat above them to relay the data they are collecting on the seabed. On other dive sites, we use a surface-supplied diving system, meaning that the diver is physically attached to the boat via air hose. In those cases, we use a hardwired system. At Las Aves we would be using scuba, not connected in any way to the dive boat, which would be some distance away. For that, we would need a wireless system.

Our communications gear for that expedition consisted of cordless microphones and headphones. Specifically we used the Aquacom SSB-2010 transceiver. It's a three-watt multichannel single-sideband underwater telephone designed with professional search and rescue teams in mind. The SSB-2010 can be used with virtually any style full face mask (ffm) or mouth mask and can also work as a portable radio on the surface. It has a transceiver box that we usually mount on the tank strap or buoyancy compensator. It feeds to our full face mask and has a special water-resistant microphone and two earphones that attach on the mask strap. With this the divers can transmit to the diving supervisor and archaeologist on the surface.

Headphones are called "bone phones" because the earpieces in these specialized headphones just touch against your head and transmit sound through bone conduction.

The mics and bone phones were essential, but of course they would not work if the diver could not talk because of a regulator in his mouth. A regulator is the device that controls the air coming from the tank to the diver's mouth. To get around that problem, the primary diving system that we have used for many years is the Aga, by a Swedish company called Interspiro.

The Aga system consists of a clear, fully enclosed face mask that looks like the masks explosives experts wear when they are dismantling a bomb. Instead of the traditional mouthpiece the diver holds in his teeth to breathe, the Aga has a cup called an oral-nasal mask that goes over the nose and mouth, like an oxygen mask, or the mask a fighter pilot wears over his face. The oral-nasal mask clears water out of the face mask automatically. The Aga was first developed for firefighters using supplemental air who needed to talk to one another. It was later modified for divers.

Aga is popular among commercial divers and is also used by police departments. The New York City police like it because the diver's face is not exposed, which helps when working in contaminated water, like the East River. Aga is good in very cold water, too. Those were not our concerns, of course. We would be diving in pristine water the temperature of a warm bath, but the oral-nasal mask also keeps the diver's mouth free and allows for a two-way communications system. Since it is dry inside the mask, we can mount a mic in there. Underwater archaeology was considerably more difficult before Aga was developed. When we first used the system on the *Whydah* project it was revolutionary, but now it is fairly common.

The diving gear would get us down to the wrecks, and the com gear would allow us to talk. We also needed the tools necessary to do the job of finding and mapping wrecks we had set out to do. Over the years, we have used all types of archaeological equipment in our search for shipwrecks, some of it common, some esoteric, some we developed ourselves.

We coined a scientific term for one tool, the "aqua-probe." A journalist who came out with us to the *Whydah* site was intrigued. She kept hearing us use the term and wanted to know how it worked, who had developed it, and how it was used. She was disappointed, I think, to find out that the aqua-probe was just a piece of rebar sharpened at the end, which we would push into the sand to see if there was anything solid underneath. Simple and effective, but it had to have a fancy name.

One of the best tools used in searching for shipwrecks is the magnetometer, which we had brought on the first trip to Las Aves but had never used, since we could never get a boat out over the reef.

Normally, the magnetometer would be towed behind a boat, using satellite navigation to set up a grid on which to steer. The person coordinating the search uses the satellite navigation equipment to determine a course to start the search pattern. He says something like "Okay, steer this line for five hundred meters and then come back," and the boat is steered on the course indicated, towing the magnetometer behind. We call this "mowing the lawn."

Doing this over and over sets up an overlapping pattern covering the ocean bottom. The magnetometer reads the magnetic fields of the earth, and it can pick up magnetic disturbances caused by the presence of ferrous metal. For our work, the metal is often iron in cannons or anchors, the largest artifacts generally left at the wreck site of a wooden ship.

The magnetometer is usually an essential tool, but I chose not to bring it on our second trip. Having seen the site and read the history of the wrecks, I did not think we could "mow the lawn" in the surf line. D'Estrées' fleet went up on the reefs, and that was where we would find them.

I probably would have brought it anyway, given my practice of bringing everything. But there was another problem. I had received several phone calls from Venezuelan officials asking specifically about the magnetometer. I didn't know why, but I didn't like the sound of things. The last straw was an e-mail from Charles Brewer. Apparently, they were asking him a lot of questions about it, too.

> I am going through a very difficult situation with the military authorities here. I know it sounds like Cuba or Ruanda-Burundy, and it is!

Frankly, I was concerned that the magnetometer would be confiscated as "a potentially subversive device" if we brought it. I opted to leave it in Provincetown.

We did bring two other types of metal detectors—both have either headphones or bone phones that buzz when metal is within a few inches. One type is the kind that people commonly use on the beach. They are used to make a general sweep of an area. We use White's of New England. Although they are relatively inexpensive, they have never failed us.

The other type is called a probe. It consists of a straight shaft, about one foot long, coming out of the control box that the diver carries in his hand. This is used to pinpoint exactly where a metal object is buried. The probe is an essential piece of gear, but it has not been manufactured to withstand the rigors of commercial work and frequently breaks down. More than once, I've wanted to throw probes into the sea in frustration. We always end up bringing four or five detectors with us.

Tape measures and underwater writing pads, along with a very sophisticated Trimble satellite navigation system to use in determining the exact location of the wrecks, completed the equipment we had brought.

We had a simple mission statement for this expedition. There would be no excavation, no disturbance of any of the artifacts we found. Our plan was to identify as many wrecks as we could, using d'Estrées' map, and to bring a little better technology to map the

Diver scanning a coral formation

site. The BBC would also be making a documentary about the project.

All our plans—the museum, the conservation center, the training program for locals—they would be for the future. We had two weeks to do what we could. I told Todd that he was going back to boot camp.

17

The Recidivism of Thomas Paine

Such pirates you will exterminate so far as in you lies, as a race of
evildoers and enemies of mankind. . . .
—*King Charles II to the Governor and Magistrates*
of Massachusetts

SPRING 1683
ST. AUGUSTINE, FLORIDA

Paine left Jamaica with his commission from Lynch and his bark, *Pearl*, of eight guns and sixty men, sometime in late 1682 or early 1683. In March 1683 he arrived in the Bahamas. Instead of apprehending pirates, however, he began to search for the wreck of a Spanish galleon that had gone down in those waters with a fortune in silver.

"Wrecking" was a favorite occupation of many who sought their fortunes at sea in the seventeenth century. It afforded the freedom of the pirate's life, with less danger and with considerably less chance of being hung if caught. Nonetheless it was the first step toward full-fledged piracy for many men, including "Black Sam" Bellamy, captain of the *Whydah*.

As it turned out, Paine never did fish for silver, at least not on that outing. *Pearl* arrived in the Bahamas to find four other pirate vessels that had rendezvoused there for the same purpose. Three were commanded by Captains Conway Woollerley, John Markham, and Jan Corneliszoon. They must have been small-time pirates, as few records of them prior to that meeting exist. The fourth ship was commanded by a Frenchman known as Bréha, possibly a nickname for Michiel Andrieszoon, whose story will come later. Together they decided that there were more profitable ventures in which to be involved.

Paine later claimed that his commission from Lynch had considerable leeway in it. In the spring of 1683, Paine and his newfound consorts decided that it could be stretched from pirate hunting to piratical raids on Spanish settlements.

It is unlikely that Lynch had such activities in mind when he gave Paine permission to "seize, kill and destroy pirates,"[1] but Paine and the others chose to interpret the commission that way. They decided that nearby St. Augustine, Florida, was a fine place to start.

THE MOVE ON ST. AUGUSTINE

Plans to sack St. Augustine had apparently been circulating in the pirate community for some time. In January 1683, a Spanish privateer had caught a vessel near Havana that had come from the Bahamas to fish a wreck. From the captured wreckers the Spanish authorities learned they had planned to move on St. Augustine after working that wreck. The Cuban authorities warned the governor of St. Augustine, Juan Márquez Cabrera, of the impending attack, and the governor tried his best to make preparations.

The walls of the city and the defensive works, the Castillo de San Marcos, were woefully inadequate. Márquez drove the men of St. Augustine to repair and build up the castillo. He even petitioned the church for permission to make the men work on holy days, but the church officials refused, in part because they did not like the governor. It was only after the governor went over their heads that he received permission.

Along with strengthening the walls of the city, the governor had two watchtowers built, one above the town and the other about twenty-one miles away at the water approaches. These were to warn St. Augustine of the pirates' approach.

Paine and the others were making preparations of their own. On the English island of New Providence, where the modern resort town of Nassau is located, they found an old native of St. Augustine, one Alonso de Avecilla. The pirates attempted to secure Avecilla as pilot and guide for their mission. When Avecilla sought refuge in the house of a Quaker on the island, the pirates, interestingly enough, got permission from the governor to seize him. The governor might not have been aware of the pending attack on St. Augustine, but if so, he was probably the only person in New Providence who didn't know.

The pirates' plan was simple. They would arrive off the Florida coast with one large vessel and three smaller ones. From there, a majority of the freebooters would take to smaller canoes, known as *pirogues,* and work their way up the Matanzas River, taking the watchtower and any Spanish outposts along the way.

On March 29, 1683, the flotilla arrived at Matanzas Inlet, south of St. Augustine. They disembarked, took to their *pirogues,* and made their way upriver in the dark, passing the lower watchtower and then coming ashore above it. The pirates hid themselves on the beach and made a stealthy approach on the tower at dawn.

As it happened, the Spanish troops stationed in the tower were all asleep. The pirates took them without a fight.

Paine's men tied them up and continued on their way, taking with them one of the Spanish prisoners to act as a guide on the river. The guide deliberately strayed, slowing the pirates' approach.

Paine and the rest were quickly losing any chance for surprise. The sentinels in the second tower spotted them and abandoned their post to spread the alarm. They met a man on horseback who brought word quickly to the governor at St. Augustine. At the same time, a soldier passing near the Matanzas River spotted the pirates, and swimming to safety, further spread the alarm.

Thus warned of the pending attack, Governor Márquez had the civilians brought inside the still-unfinished walls of the castillo, where they found some degree of security, but no facilities for housing or cooking. The governor mobilized all of the regular troops as well as the militia and dispatched two small units to lie in ambush for the pirates before they reached the city. He also posted lookouts at the outskirts of the city to alert him of the pirates' approach.

With that accomplished by late afternoon, Governor Márquez turned his attention to the castillo's walls. With considerable energy and a decided lack of diplomacy, the governor drove the people inside the city to work, cursing and swearing as they labored.

Church officials claimed that Márquez's behavior was so offensive that the people's fear of the pirates was soon replaced with anger at their governor, and only through the intervention of a priest was a mutiny avoided.

In the end, the walls were never needed. Governor Márquez showed considerable skill in defending the approaches to the city, and Paine and his consorts did not display the determination of a de Grammont. On the morning of the 31st, the scouts reported to Márquez that about forty pirates were approaching the city. The governor dispatched another ambush party that hid a mile or so southeast of the castillo.

The pirates in fact numbered around two hundred men, but they weren't expecting the initiative displayed by Márquez. They marched boldly into the ambush and were greeted with a devastating musket volley. After a brisk firefight, the pirates fell back, thwarted in their advance a mile from the city. The only thing they had to show for their efforts was a prisoner, private Francisco Ruíz, but even he would ultimately contribute to their failure.

The pirates retreated some distance and halted to reevaluate their position. They tortured Ruíz to discover the strength of the city, but Ruíz fed them disinformation. He claimed the castillo was ready for them and well manned, and that there were ambushes laid out along the way. The governor had rounded up all available carpenters to build carriages for the castillo's cannons and to fully repair the walls.

The pirates then turned on their unwilling guide, Alonso de Avecilla, and threatened to kill him for not telling them about the new watchtower, but Ruíz interceded. He told the pirates that Alonso could not have known about the tower, since it was built after his departure from St. Augustine. At last, the pirates made Ruíz agree to guide them back to the city. This Ruíz agreed to do, but he told them that there were no guarantees, as the governor had set ambushes at unknown locations along their route.

That was too much for the pirates. They abandoned their overland attack on St. Augustine and retreated back to the captured watchtower at Matanzas Inlet, dragging the unfortunate Ruíz with them. It would be another two and a half years before Ruíz was back in Spanish territory as a free man.

For three days, Paine and the pirate captains remained at the watch-tower debating what they should do next. Paine was not a man to shed blood lightly. At last, on April 5, they brought their ships into St. Augustine inlet with the intention of taking the city from that direction. Once again, they got cold feet, daunted by the sight of the Castillo de San Marcos and recalling the reports that Ruíz had given of the castle's preparedness. They decided that St. Augustine could not be taken, boarded their ships, and sailed away, with nothing but casualties to show for their efforts. The attempted sack of St. Augustine was one of the last of the pirate land raids in the territory of what would become the United States.

The buccaneers sailed north along the coast, sacking a few smaller towns on the St. Johns River and Amelia Island. Stopping off at what is now Cumberland Island off the coast of Georgia, they careened their ships and buried the men who had died of wounds received in the ambush. There they released the prisoners they had taken, except for Private Ruíz.

Paine, Markham, and Bréha returned to New Providence Island, but there was no welcome for them there. The governor, who had given them leave to capture Alonso de Avecilla, now professed every intention of arresting them, but could not muster enough men for the task.

Paine returned to the wreck site he had initially intended to fish, but found that there were many others now working it. Later, when a large ship sailed into New Providence, the governor manned it with sufficient force to overcome Paine and dispatched it to the wreck site. By then, Paine and the rest were gone.

Thomas Paine's activities, and the friction thereby caused between England and Spain, caught the attention of the very highest levels of officialdom. About a year after the St. Augustine raid, no lesser figure than King Charles II of England wrote to the governor and magistrates of Massachusetts:

In consequence of the Ravages of pirates in the territory of the King of Spain, we have thought fit, for the encouragement of the amity that exists between us and his Spanish Majesty, to give orders for the suppression of the pirates, and that you give no succor nor assistance to any, and especially not to one called Thomas Pain, who with five vessels under the command of

The Captain

Captain Breha, has lately sailed to Florida. Such pirates you will exterminate so far as in you lies, as a race of evildoers and enemies of mankind. . . . [2]

The awesome power of the king notwithstanding, it would take more than the colonial governors could muster to bring down Thomas Paine. If nothing else, Thomas Paine was a survivor.

18

A Homecoming
for a Pirate

Oh sweet it was in Avès to hear the landward breeze,
A-swing with good tobacco in a net between the trees,
With a negro lass to fan you, while you listened to the roar
Of the breakers on the reef outside, that never touched the shore.
—"THE LAST BUCCANEER"
Charles Kingsley

SUMMER 1683
NEWPORT, RHODE ISLAND

Though the English, French, and Spanish authorities routinely alternated between tolerating piracy and accusing one another of tolerating piracy, the English governors began to turn on one another as well. And with good reason. The leaders of the various colonies had very different levels of tolerance when it came to piracy. Sir Thomas Lynch had already complained about the amount of cooperation the pirates received in the North American colonies. After Paine used Lynch's commission to launch his attack on St. Augustine, the Earl of Craven fired back:

I have read what Sir Thomas Lynch has written you about the reception of privateers at Carolina. . . . At Providence, which Sir

T. Lynch has complained of before now for harboring pirates, all imaginable care was taken to suppress them, and no attempt upon the Spaniards was made except by the instigation of a person whom Sir Thomas Lynch had sent to take pirates.[1]

This was not the last time that Paine would be at the center of a storm.

In the fall of 1683, Paine returned to Rhode Island. His name was well known there, as were his questionable activities of the past years. Government officials called for his arrest. Fortunately for Paine, the governor of Rhode Island, William Coddington, was not one of them. In fact, for unknown reasons, Coddington did everything he could to obstruct efforts to bring Thomas Paine to account.

One of the governors who called for Paine's arrest was Edward Cranfield of New Hampshire. In October 1683, he related that Paine had come in "with a counterfeit commission from Sir Thomas Lynch styling him [Lynch] one of the Gentlemen of the King's Bedchamber, instead of his Privy Chamber, whereby I knew it to be forged. Colonel Dongan [governor of New York] and I asked the government to arrest [him], but they refused."[2] A few months later, Paine's ship *Pearl* was briefly detained in Boston but was soon released. For the moment, Paine was safe.

The next spring, Paine's past once again came back to haunt him. Just when it seemed royal officials were losing interest in prosecuting Paine, the declaration from the king of England arrived, specifically naming him as one of the worst offenders of that "race of evildoers" to be exterminated.

The deputy tax collector of Boston, T. Thacker, attempted to impound Paine's ship at Newport and have Paine arrested. Like any government official, Thacker filed an extensive report:

By seven or eight at night I had satisfied myself as to the character of the ship, waited on Governor William Coddington, and shewed him my Commission and demanded his assistance in seizing her. He put me off, promising to answer me next morning, by which time the pirates had time to arm themselves against arrest.[3]

Thacker does not hesitate to call Paine a pirate. Rather, it was the governor who equivocated, and would continue to do so. Like Governor Dongan of New York, Governor Cranfield of New Hampshire was also in Newport, and they too joined in the fray. Cranfield had already called for Paine's arrest the year before, and still wanted to see the pirate locked up. Thacker continues:

> I went to the governor the next morning, but instead of giving assistance he avouched her [Paine's ship] a free bottom as having a commission from Sir Thomas Lynch. . . . I asked to see it, and it was presented by Paine, in the presence of Governor Dongan and Cranfield of New Hampshire and others. It appeared to be a forgery, Governor Cranfield and others affirming that it was not Sir Thomas Lynch's hand, nor were his titles correctly given, but Governor Coddington was of other mind and declared her a free bottom.[4]

The next day, Thacker continued to urge the governor to seize the ship and men, "especially Thomas Paine, as the Commission was certainly false, and the ship had not been to Jamaica but on a piratical cruise, and had plundered the town of St. Augustine," giving Paine more credit than he was due.

Coddington continued to refuse to arrest Paine and informed Thacker that if he wished, he could prosecute Paine through the courts. Thacker said he would be happy to, if Coddington would arrest him. Not only did Coddington refuse to do so, he would not supply Thacker with a copy of Paine's commission from Sir Thomas Lynch.

As soon as Thacker returned to Boston, he tried one last time to convince Coddington that Paine's commission was a fake by sending off a sample of Lynch's handwriting to the governor. In frustration he later reported, he ". . . sent him [Coddington] one of Sir Thomas Lynch's passes to convince him, but he would not see with eyes like other men."[5]

Why Coddington went to such lengths to protect Thomas Paine is something of a mystery. It might have been rivalry. Coddington and Cranfield had their own problems. Just days after the affair with Thomas Paine, the two men engaged in a bitter and acrimonious fight concerning unsettled claims of land and jurisdiction in the Narragansett area, with Cranfield at the head of a royal commission looking into the matter.

It is entirely possible that this dispute had already started, and the land question was the reason that Cranfield was in Newport at the time. Perhaps Coddington was feeling protective of his fellow Rhode Islander, especially in light of the fact that the two governors and the tax collector were representatives of the Crown, New Hampshire and New York being Crown colonies. Coddington might have felt it his duty to protect his fellow Rhode Islander against such tyranny. It was the attitude of independence that would lead to the American Revolution nearly a century later.

Some of the problems may have stemmed from the character of Rhode Island in those days. Unlike other areas of New England, Rhode Island was a uniquely liberal and tolerant colony, a freewheeling little place, founded by religious dissenters and iconoclasts. Nor was it a particularly wealthy colony; enough money distributed in the right directions could generally quash curiosity as to the origins of a man's personal fortune. In 1657, a Dutchman writing from "New Amsterdam" (modern New York) noted that Rhode Island was "the Receptacle of all sorts of riff-raff people, and is nothing else than the latrina of New England. All the cranks of New England retire thither." While this observer was certainly exaggerating, Rhode Island was a nursery for smugglers and pirates during the 1680s and 1690s.

In any event, the chief legal point was the validity of Paine's commission. Though it is unlikely that Sir Thomas Lynch intended to give Paine permission to sack St. Augustine, it does appear that he gave the filibuster some sort of commission. That much is indicated in his letter to Sir Leoline Jenkins, in which he states he "accepted the offer"[6] to grant Paine a commission to hunt pirates.

Knowing that Paine went to Florida on the strength of the commission that he, Lynch, had issued, Lynch must have been shaken to read the king's proclamation, stating, "You will permit no succor nor retreat to be given to any pirates, least of all to Thomas Pain, who . . . is lately arrived in Florida."[7] Perhaps this is why little evidence exists concerning Paine's commission, beyond Lynch's letter to Jenkins.

The fact is that Cranfield was right in his assessment of the commission that Paine produced. Lynch was not a Gentleman of the King's Bedchamber, as the commission apparently stated, but was in fact a Gentleman of the Privy Council. The Bedchamber was reserved for peers of the realm, not men with mere knighthoods. Confusing the two was not a mistake that Lynch would have made.[8] Paine had shown the governors a forged document.

So what was the truth behind Paine's commission? If Lynch was

prepared to issue Paine a commission, why did Paine end up with a
forgery? Perhaps Lynch changed his mind before actually issuing the
document, leaving Paine to write his own. Perhaps Paine lost the
original and tried to re-create it from memory.

Perhaps Lynch was not so innocent in the affair as he should have
been. It is worth noting that he was posthumously accused of irregular
practices in connection with a later pirate raid.

One of the perpetrators died at Port Royal shortly after the raid, and
his effects, including treasure, were seized by the government. While
most were forwarded home to London, as was right and proper, the
attorney for the pirate's widow claimed that a parcel of "Spanish Jew-
ells" somehow managed to end up in the possession of Governor
Lynch's wife. It was also alleged that the wife of the Jamaican admi-
ralty agent had received a ring from a Dutch freebooter, and that
Lynch had taken bribes from assorted French pirates who were fearful
of returning to Petit Goâve and sought asylum at Port Royal.

We know only that Thomas Paine had a piece of paper with the
name Sir Thomas Lynch on it. He used it to justify his attack on St.
Augustine, and he used it later to avoid the possible repercussions of
his actions. It was authentic-looking enough to satisfy Governor Coddington, though Coddington was apparently eager to let Paine off.
Wherever it originated, Paine certainly got a lot of mileage out of that
document.

Despite Coddington's protection, Paine remained the subject of
official badgering and Coddington remained under pressure for pro-
tecting him. In September, royal agent William Dyer wrote, "I have
also caused Thomas Paine the arch-pirate, to be secured, and charged
the Governor of Rhode Island with him and with his own neglect for
not assisting the Deputy Collector to seize him and his ship."[9] Once
again, however, Paine somehow managed to avoid prosecution and
gave the gallows the slip. Thomas Paine had a lot more fight left in
him.

19

In the Wake of Jean Comte d'Estrées

October 26, 1998
Las Aves

After setting up and testing all of our gear in the Blue Room, we packed it all up again and loaded it on a plane, this time a small island-hopper that took us to Los Roques, the nearest inhabited land to Las Aves. That is not to say it is densely populated. Los Roques is a smattering of low brown-and-green islands that seem to float in the light-bluish-green water. It was once a hangout for my old friend Captain Sam Bellamy of the *Whydah*. We landed on an airstrip that looked like a straight gray scar across one of the larger islands, running from one shore to the other.

There we met our dive boat, the *Antares,* an eighty-five-foot cruiser that would be our home and work platform for the next two weeks. The *Antares* is a PADI dive boat. (PADI stands for Professional Association of Diving Instructors. It is a parent organization for sports diving instructors.) PADI rates dive boats like hotels, and the *Antares* was rated five stars. By the standards of American or European hotels, she might not have been five stars, but compared to the *Vast Explorer,* the boat we used for the *Whydah* project, a no-frills workboat, the *Antares* was the height of luxury.

The *Antares* was a big, boxy vessel. Her hull and deckhouses were white. The lower deck was made up of cabins. The next deck, at the

Antares returning to Los Roques

level of the open afterdeck, included a big salon and galley. The salon was decorated in the unfortunate earth tones of the mid–1970s, but what it lacked in taste it made up for in roominess and light. It had a wide parquet floor over which we were strictly warned we could not drag dive equipment or even walk with shoes. The salon became the central gathering place for the entire expedition. We met there, ate there, partied there, and argued there.

The *Antares* was set up as a dive boat, and that made her ideal. Carrying divers was her purpose in life, so she had on board everything we needed for that part of the expedition. She had air compressors and storage for tanks and gear. Todd Murphy had carefully figured the number of air tanks we would need, and the captain of the *Antares* had seen to it that they were all aboard.

The captain was Ron Hoogesteyn. In his late thirties, he is an imposing figure at six foot two and 260 pounds. He is from a well-known Dutch family in Venezuela. A diver himself, Ron understood what was needed for our expedition and he made certain that his boat was fully equipped.

I liked Ron right off and really came to appreciate his understanding of Las Aves, everything from the flora and fauna to the protocol of

Captain Ron Hoogesteyn (third from left) with Barry Clifford and crew

dealing with the coast guard. Ron's love for the environment at Las Aves is evident, and he helped me appreciate what a beautiful and fragile ecosystem exists there.

Before setting out, we discussed some of the potential problems we might encounter. Ron knew Las Aves, having taken sports divers out there, and he knew that there might be problems with big surf and difficulty getting to the dive sites. He knew we might face other problems as well—sharks. Las Aves is a favorite hangout for lone hammerheads. Hammerheads are most dangerous when they are by themselves.

"You have tigers down here, too, don't you?" I asked.

"Yeah. Nasty ones," Ron said, by way of encouragement.

Later in the expedition, when my own spirits were low and I was sick as a dog, Ron proved to be so eager to help it was irritating. If that is the worst problem a guy presents, then that's fine with me. We were lucky to have him.

The next morning we were under way, steering west toward Las Aves, plowing the sixty miles to the atoll through that beautiful blue-green ocean with the trade winds at our back. We were all aboard and the spacious afterdeck was crammed with our gear. It was the last leg of the trip to the site, and we could all feel the excitement. The weather was fine but the seas were up and the trip was somewhat reminiscent of the bone-jarring ride we had had months before on our first boat ride to Las Aves. But despite that, we were all in good spirits for the upcoming explorations.

It was interesting to think that we were sailing close to the same course d'Estrées had been sailing on that terrible night more than three hundred years before. We, of course, enjoyed many advantages that d'Estrées did not. Foremost among them, we knew that the reefs were there.

As we approached Las Aves, we saw that we were not the only ones who knew about the reefs and the potential wealth they held. Another vessel, about the size of the *Antares,* was also making for Las Aves. As I mentioned before, there are fishing boats that go out to the island, and the occasional sports diver. There is also a Venezuelan coast guard base. But this boat did not look as if it was any of the above.

Ron and I took turns examining the stranger through the binoculars as she and the *Antares* drew closer.

I asked Ron, "Do you know what boat that is?"

Ron did not. He said, "We heard them on the radio, talking to the navy. They are going to dive at Las Aves."

Everything about her—the size, the type of vessel, her apparent destination—suggested that she was heading for the reefs for a little treasure hunting, or was I being paranoid?

"This is a treasure-diving boat," I said. "They're treasure hunters."

Once word of an old shipwreck gets out, it can set off a gold rush. The ones who usually show up are the amateur treasure hunters who have never found anything of their own. They arrive off your site like the wild dogs of Africa, sniffing the air and circling, waiting for a chance to grab a few scraps and run. It was like that with *Whydah* and other projects. Never mind the fact that the wrecks at Las Aves probably did not have anything that would interest a real treasure hunter.

We never found out for certain who was on that other boat. My best guess is that as we were checking them out, they were watching us. They must have figured out who we were. Long before we came up with one another, they sheered off and headed for the horizon. We never saw them again.

Still, there were good reasons to be cautious. Along with treasure hunters, there is a real threat of piracy in the Caribbean. Not all the buccaneers died in the eighteenth century. While we were at Las Aves we got a report of a yacht found floating and vacant. There was no clue as to what had happened to the crew, but one can guess. Every now and then during our stay at Las Aves we would see a beat-up boat motor slowly by, checking us out, and we'd think, Here we go . . . But we were never bothered.

Ron took the *Antares* carefully around the reefs and into the lagoon.

It is beautiful. The clear water of the lagoon is aquamarine, hedged in a thick tangle of mangrove, punctured by a small black stream that flows into the heart of the island. After the anchor is set, I go to the dive platform at the stern. The air is heavy with dampness, and the smell of the swamp is thick and syrupy.

Just a week before we had been on the Cape, bound in woolens against the cold late-autumn wind. The sun, when it appears, hangs barely above eye level. Here, the sun is directly overhead, just twelve degrees north of the equator.

I have been traveling for seventy-two hours. I am tired, dirty, and my skin seems to crawl with imaginary insects.

I am out of my filthy cargo pants in less than a heartbeat and stand naked on the platform for a moment. I look like one of those Bulgarian men you've seen in *Life* magazine, who bathe in the Black Sea in winter—white as a polar bear with a layer of winter fat.

I step off the platform and let myself sink to the bottom of the lagoon. I sit on a mat of pulverized pink coral; taking a handful, I begin to scrub myself from head to toe. Looking up, the round belly of the *Antares* appears like a June bug floating on air . . . except I cannot breathe here.

I let go of all my senses and swim for the mangroves about a hundred feet away. The tidal flow of the sea carries me into the swamp. My lungs begin to ache. I break the surface, letting my air out slowly, then ride the flow of the warm seawater toward the heart of the island. The bull shark sometimes hunts here. . . . I am still as death, but never have been more alive. I think we all had an unspoken sense that we had to dive into the tropical sea to wash ourselves clean of the winter cobwebs, to literally immerse ourselves in the beauty of the lagoon.

My team was aboard the *Antares,* the divers and support people who had come with me, as well as Mike Rossiter and the BBC crew. Max and his friends had chartered a large sailboat—very well appointed and a bit more luxurious than the overcrowded *Antares.* They were anchored nearby.

Charles had his own boat and his own team, including a film crew.

Mike Rossiter, Alan Barker, Todd Murphy, and Guillermo Cisneros

His cameraman was Guillermo Cisneros, the son of industrialist Gus‐
tavo Cisneros, reportedly the wealthiest man in South America.
Guillermo had chartered the boat for Charles's team, a big power
yacht with all the amenities. Still, as cushy as his boat was, Charles
spent most of his time on board the *Antares,* the heart of the operation,
central control.

Charles had plans for his own documentary, which worried me.
Afraid of conflict with the BBC/Discovery contract, I chastised myself
for not having made Charles sign something before we showed up
there. Mike was not concerned, however, and since Mike was the
BBC producer, I tried to be unconcerned as well.

It was early evening when we anchored in the lagoon, with the sun
sinking toward the western horizon and lighting up the little hump of
land that was the island of Las Aves yellow and gold. It was too late to
get started; we settled for preparing for an early dive the next morn‐
ing. The team checked over the gear we would need, saw that it was
assembled and in working order. Todd Murphy, Charles, and I pored
over the charts, ancient and modern, and discussed where best to start
looking for wrecks.

Charles and I had different ideas as to what the mission should entail. He wanted to choose just one wreck and partially excavate it. I vehemently opposed this idea as we had neither the personnel nor the time to conduct a proper excavation—much less the artifact conservation such a project would entail. Not to mention the fact that our plans and permits were predicated on filming only.

Charles and I had not butted heads for long before I realized that I just had to focus on what I was doing and let him do what he wanted. He had an agenda, whatever it was, and I did not have the time or energy to fight with him. I was there to locate and map shipwrecks. That was all I had to worry about. I decided I would not get dragged into Charles's game. And I did not. For the most part.

We had only two weeks to dive on the site. The BBC and the Discovery Channel were chartering the boat, and that was the amount of time that they calculated they needed, so that was the amount of time we had. Sure, I would have loved to have had more, but those decisions were above my pay grade.

Also, I had a bad feeling that the authorities would be showing up any day to see what we were doing. We had permits, of course, plenty of different permits from plenty of different people, but in a place like Venezuela that is not necessarily a guarantee of anything. We knew of at least one other group with their eyes on Las Aves, and God knew how many more there might be, and what governmental strings they might be attached to. It was possible that we would be shut down before we were finished.

With that in mind, I wanted to do as much as we conceivably could as quickly as we could.

20

A Visit from the Navy

There is a set series of steps one takes in underwater archaeology (or any kind of archaeology, for that matter). First, a great deal of research is done, in order either to find a site or, if the location is already known, to find out what happened there, to anticipate what you are likely to uncover and what you should keep your eyes open for. A historical context needs to be established.

Along with the research, you have to prepare documentation and make a plan outlining what you hope to accomplish at the site, be it survey and mapping or recovery of artifacts. You plan as best you can, but of course you cannot cover every contingency until you find the wreck and see for certain what is possible and what is not. You may draw up a plan that involves recovering artifacts, but if you don't find any artifacts, that's that.

Once you are on the site, you proceed with remote sensing, using the magnetometer, metal detectors, and other sensing tools. Remote sensing will, ideally, allow you to locate the wreck, which you then inspect visually to identify and confirm what you have found.

On a site where you are planning excavation, the next step is to dig a test pit. Test pits can be of any size, but are usually around two to six feet square. Stakes are driven into the four corners of the pit, and excavation is done within the confines of that square. Carefully placed test pits will give you a sense of the concentration of artifacts, and from there, a full-scale excavation can be designed.

As the artifacts are revealed, they are carefully mapped out and detailed drawings are made and notes taken of exactly where they were found, in what position, at what depth, etc. In that way, anyone studying the site later will always be able to see how the artifacts were found, in situ, before removal. A lot can be learned from that information.

Then, with mapping and drawing done, the artifacts are carefully removed. Since most objects that have been in salt water for centuries will start to break down when they hit the air, conservation of artifacts begins immediately, right on the dive boat, even before they can be transported to the conservation laboratory where long-term conservation is undertaken.

We knew that we would not be excavating or removing artifacts. Our permits were for filming only. With that caveat, the most useful thing we could do, from an archaeological standpoint, was to document everything, map every wreck we could find, make careful diagrams of each site, and pinpoint each site on the map. We could document the direction each ship was moving when it ran aground, distribution patterns of the artifacts, how much material has been shifted around over the years, and scatter patterns for various types of artifacts.

It would not be easy, but I figured that even with only two weeks we could do that for every site on the reef, at least every shipwreck that d'Estrées had marked. We hoped that the BBC's documentary would tell the story of d'Estrées and his buccaneer mercenaries and of the discovery and exploration of those wrecks, and illustrate how modern underwater archaeological mapping is done.

I was eager to test the accuracy of d'Estrées' map. I had overlaid what the French admiral had drawn with a modern aerial photograph of Las Aves to see if the wrecks really were where they should be. This was partially my curiosity but also an aid to future explorations. If you know exactly where to find a wreck, that can save a lot of time in the archaeological process.

Lastly, there were those pirate ships, the *flibustiers*. Even if we weren't excavating, I wanted to find them.

We knew what we wanted to do, and we had the team and equipment to get it done. We went to bed that night, lulled to sleep by the gentle rocking of the *Antares* in the sheltered water of the lagoon.

The next morning the sun rose to reveal the beautiful blue-green water, the sharp horizon where sea met cloudless sky. Margot and I got up early and went for a swim. It is something we try to do every morning when we can, especially on Cape Cod in the summer. At Las Aves it was wonderful, the water clear and milky warm. We swam up to the mangroves, reveled in the beauty of the place. Margot and I had not been together very long at that point, and it was as romantic as it was beautiful.

That glow might have lasted all day if we had not returned to the *Antares* to find a Venezuelan navy ship anchored not half a mile away.

Oh, great, I thought. We hadn't even put on our wet suits.

Still, there was no real reason to worry. Antonio had secured all the permits we required, and even some we probably did not need. Charles had secured even more—Mike had videotaped him doing so.

The naval vessel was not huge, more of a gunboat, somewhere around 150 feet, but it was intimidating enough. As a boat put off from the gray ship and headed straight for the *Antares,* we assembled our permits, passports, and sundry paperwork.

We welcomed the naval officers aboard with courtesy, and they returned our greetings in every way. They were pleasant, friendly, and professional in their crisp khaki uniforms. They had a job to do and they went about doing it in a businesslike manner.

It was not a coincidence that they were there. They had come out specifically to look into what we were doing. The officer in charge knew the names of everyone aboard, even as he called to see our passports. He obviously had been given the paperwork that we had submitted to the Venezuelan government during the permit application process.

We lined our people up with passports in hand, and one by one they were checked against the navy's list. Everything was in order.

Almost.

It was sort of a tricky situation with Todd Murphy and Carl Tiska, a member of the U.S. Army Special Forces and a high-ranking U.S. Navy SEAL on the same boat. We joked that the Venezuelans might think we were planning an invasion. That was before we actually had the Venezuelan navy on board. Though both men were with us purely for archaeological work, we were concerned about how it would look, what suspicions it might arouse.

Todd, however, does not like to advertise his presence and his military connections. He had not submitted any paperwork to the

Venezuelan government that mentioned his status in the Special Forces. They never asked.

I knew that Todd and Carl took this stuff seriously. They joked about an invasion, but they were also making preparations. Between them, from force of habit, they devised an escape route and contingency plans in case things got ugly. Being prepared for any eventuality was something their military training had ingrained in them.

Our crew passed muster, and the navy moved on to the permits. Charles was the only member of our team who spoke fluent Spanish. He and the officer from the Venezuelan navy went round and round, while we all watched and tried to guess what was happening. It was like a wolf counting sheep.

Finally, Charles explained what had transpired. Despite the reams of paper we showed him, the naval officer had decided that we did not have the right permits. He was shutting us down.

I couldn't believe it was happening.

Hoping to console us, I suspect, the officer assured us that he was not stopping us completely. We could still dive; we just couldn't film anything. Considering that a major reason that we were there was to shoot a documentary, that concession was as good as worthless.

We tried to be as persuasive as we could. We had filming permits, which we showed him, and expedition permits and God knows what other kinds of permits, but he would not be swayed. We could not film until we had the proper papers, which he did not see among the many permits we showed him. We were dead in the water.

The navy men bid us a friendly good day and returned to their ship. They did not pull up anchor and leave, as we had hoped, but remained conspicuously in place, just off the coast guard station, watching.

21

"A Great and Mischievous Pirate"

1679
The Spanish Main

The Greatest of Them All

Of all the filibusters cast up on the beach at Las Aves, the one destined for the greatest piratical career was the Dutch-born Laurens Baldran, known to the Brethren of the Coast as Laurens de Griffe, or, more commonly, de Graff.

De Graff was a born leader of men, fearless and at the same time reportedly refined and genteel. It was said that he "always carries violins and trumpets aboard with which to entertain himself and amuse others who derive pleasure from this. He is further distinguished amongst the filibusters by his courtesy and good taste."[1] Some of this may have been hyperbole prompted by literary conventions of the time, but the last part of this description would be borne out by de Graff's career.

Sir Henry Morgan's characterization of de Graff would also prove correct, when he called him "a great and mischievous pirate,"[2] which is most ironic coming from a onetime buccaneer king.

De Graff was described by the same eighteenth-century writer as being tall, blond, mustached, and handsome. At best, that description was sheer speculation, as de Graff had been dead at least twenty years when it was penned. At worst, it was a fiction crafted to hide the truth and prevent the example of de Graff's life from spreading. In point of fact, early Spanish sources indicate that Laurens de Graff was a runaway slave of African heritage. The nickname "de Griffe," by which he was often called, is an old term for a mulatto of three-quarters African ancestry. That Spanish sources might be more forthright about "Lorencillo's" background can be attributed to the fact that the French and English had far more to fear from slave revolts than did the Spanish. The last thing an eighteenth-century French writer in Haiti would want to write about was a slave who escaped and then rose to wealth and fame at the head of a band of outlaws and pirates.

Details of his early life are sketchy. It is most likely that he was born in Holland. He was captured by the Spanish in the Low Countries and enslaved sometime during the 1660s. The Spanish had a talent for creating their own worst enemies, and de Graff was one of them. His hatred of the Spanish never wavered.

At some point during his Spanish captivity, de Graff was brought to the Canary Islands. Spanish plantation slaves were allowed to wed, and de Graff married a woman named Petronila de Guzmán, who was possibly descended from Jewish refugees. He was soon separated from her, however, and put aboard a Spanish galley. The ship aboard which de Graff was forced to serve was part of a special naval squadron called the Armada de Barlovento, which was tasked to combat piracy in the Caribbean.

In the early to mid 1670s, de Graff escaped the galleys, with nothing but a burning hatred for the Spanish. Like many escaped slaves or naval deserters—he was both of these—he turned to piracy. His career lasted for three decades. In that time, he brought down havoc on the heads of the Spanish, both as a pirate and as a commissioned officer in the service of France.

Historians of the seventeenth and eighteenth centuries never mentioned de Graff's race, or, if they did, they blurred the facts. An uprising of black slaves was the worst nightmare of the eighteenth-century colonial elites, threatening not only their personal fortunes but also the economies of those European countries that were amassing more and more wealth from the West Indies.

The revolt in Haiti led by Toussaint-L'Ouverture a century later proved that the slaveholders' fears were well founded. No white men,

A seventeenth-century galley

historians included, wanted to even hint that violent resistance could garner for enslaved men the freedom, riches, and success enjoyed by Laurens de Graff. It is no accident that de Graff's name has been eclipsed by such lesser men as Henry Morgan and William Kidd.

Of all the buccaneers of the seventeenth century—L'Ollonais, de Grammont, Paine, Yankey, and the rest—the black pirate Laurens de Graff became the most successful. So great was his fame among his peers that a French historian later wrote, "When it is known he has arrived at some place, many come from all around to see with their own eyes whether 'Lorenzo' is made like other men."[3]

A PIRATE'S CAREER PATH

Spanish historians claim that de Graff's first action as a pirate captain came in March 1672. On the last day of that month, a band of pirates slipped ashore at the Mexican city of Campeche in the predawn dark. On a nearby beach, a *guarda del costa* frigate stood on the stocks, partially built, beside it a huge stockpile of lumber. This the pirates set on fire. In the light of that inferno the pirates' ships sailed into the harbor, while the arsonists already ashore infiltrated the city.

As had happened before, and would happen again all along the Spanish Main, the citizens of the city woke to find their town occupied by pirates. Realizing that they were under attack by filibusters, the defenders of the city acted as Spanish militia generally did—they fled in panic.

The next morning, a Spanish merchant ship sailed right into the harbor, unaware that the city was in the hands of the filibusters and that pirate ships lay at anchor there. Along with a rich cargo, the merchantman carried in her hold 120,000 pesos in silver, all of which the pirates liberated.

It was not a great buccaneer army that had taken Campeche, but a rather small contingent, and the pirates knew that it was only a matter of time before they would be overwhelmed. Soon after emptying the hapless merchant ship, they abandoned the city. By the time a relief column from Mérida de Yucatán arrived at Campeche, the buccaneers were long gone.

Though the Spanish claimed that the 1672 Campeche raid was in part the work of de Graff, there is no other documentation to prove it. De Graff, like the outlaws of the old west, might have received credit for more crimes than he actually committed. More to the point, no other record exists of de Graff's activities for the next five or six years, which casts some doubt on the likelihood of his involvement there.

While the details of his rise through the ranks to command are unknown, Laurens de Graff probably acquired ships in the same manner as most pirates. He started small, first commanding a modest bark and using that to capture a larger ship, and then a larger one after that. His ship was almost certainly part of d'Estrées' squadron. In the fall of 1679, a year after Maracaibo, de Graff captured a Spanish frigate of some twenty-four to twenty-eight guns, the frigate *Tigre*. Ironically, *Tigre* was a part of the Armada de Barlovento, the very squadron from which de Graff had escaped. The former slave must have been especially gratified by the capture of *that* vessel.

CRIME PAYS BIG

By 1682, de Graff was so successful a pirate that he garnered special attention from the authorities, who made special efforts to stop him. Sir Henry Morgan dispatched the frigate HMS *Norwich,* commanded by a Captain Peter Heywood (himself a future governor of Jamaica), specifically to hunt down de Graff. Morgan worried about the threat that de Graff presented, and the real possibility that de Graff might instead capture the *Norwich.* Writing to the Lords of Trade and Plantations, Morgan said:

> And that the frigate might be better able to deal with him [de Graff] and to free him [Heywood] from the danger of being worsted or taken, I have put forty good men with commanders aboard her. . . . I doubt not but your Honors will allow this charge, it being necessary for the King's service and the preservation of the frigate. . . .[4]

There is no record that Heywood ever managed to find de Graff. Instead, de Graff made one of the grandest conquests of his career.

In July 1682, near Puerto Rico, de Graff's *Tigre* intercepted the Spanish frigate *Princesa,* a fine ship the Spanish had taken from the French and incorporated into the Armada de Barlovento. *Princesa* mounted twenty-six great guns, ten smaller swivel guns called patararoes, and carried 250 men. She was very much a match for de Graff and his company.

Hollywood scenes of pirate ships battling it out with men-of-war on the high seas rarely happened in reality. The fight between *Tigre* and *Princesa* was one of those rare instances of a ship-to-ship duel, rather than a ground assault. The ships fought for hours, never grappling but rather dueling with their long guns. No doubt the Spanish captain was wisely not eager for hand-to-hand combat with the fierce buccaneers.

It was a one-sided battle all the same. When the *Princesa* struck her colors, de Graff had lost eight or nine men killed, another sixteen or seventeen wounded. The Spanish lost fifty men killed or wounded, including the captain, who was wounded in his upper thigh and had "his belly somewhat torn by a great shot from one of Laurence's quarter-deck guns."[5] Typical of de Graff's humanity, he had the wounded captain immediately put ashore, along with a surgeon and a servant to attend him.

A fight at sea

De Graff had hit the jackpot. Along with a variety of valuable goods, the ship was carrying the payroll for the garrisons at Puerto Rico and Santo Domingo, around 122,000 pesos in Peruvian silver. Symon Musgrave, an Englishman who frequented Spanish territory, reported to Governor Sir Thomas Lynch in Jamaica:

> It is said the pirates made one hundred and forty shares and shared seven hundred pieces-of-eight per man. Laurence himself is now at Petit Guavos; his ship and prize are refitting.[6]

"Petit Guavos," or Petit Goâve, rather than Tortuga, was de Graff's preferred port of call.

It is little wonder he felt comfortable there, given the official coop-

Buccaneers taking a Spanish ship

eration available. Musgrave goes on to report, "The Governor of Petit Guavos has received his share underhand but resolves to grant no more commissions. . . ." De Graff was clearly operating under some type of sanction by the governor at Petit Goâve. The Spanish, however, were being pushed to the breaking point and the governor feared reprisals. This was, after all, one of those rare windows of peace in Europe amid the almost constant warfare of the seventeenth century.

The robbery of their payroll ship infuriated the Spanish authorities. To make matters worse, it had been conducted by one of their own former slaves, who had then converted the captured *Princesa* to his new flagship. Unable to take vengeance on de Graff, Spanish authorities took it out on another freebooter of Dutch descent, Nikolaas Van

Hoorn, who was in Santo Domingo attempting to sell a cargo of slaves. In retaliation for de Graff's actions, the slaves were confiscated by the Spanish authorities. Van Hoorn managed to escape with no more than twenty of his men.

Once again, the Spanish had picked the wrong man for an enemy.

22

Nikolaas Van Hoorn

*[I]t is said that Laurens, having two good ships and four hundred
men, will not join him, and that his [Van Hoorn's] own people
and the other French abhor his drunken insolent humor.*
—*Sir Thomas Lynch*

FEBRUARY 1683
THE GULF OF HONDURAS

If Laurens de Graff represented the best of the buccaneers, Nikolaas
Van Hoorn represented the worst. In December 1681, Van Hoorn
had sailed from London in command of the ship *Mary and Martha* of
four hundred tons, forty guns, and a crew of one hundred fifty men.
He also took with him his son, who was around ten or twelve years
old. Instead of straightforward slave trading, which would have been
bad enough, Van Hoorn went on a sixteen-month rampage through
the Bay of Biscay, the Canary Islands, the African coast, and finally the
Caribbean.

Van Hoorn was a vicious drunk, brutal and utterly without regard
for the nationality or circumstance of those whom he attacked. He
was too much even for his own men. As sailors from his ship later
reported, Van Hoorn "was forced by weather into a French port in
the Bay of Biscay, where twenty-five of his men, seeing what a rogue
he was, ran away."[1] In Cádiz, he forced ashore thirty-six more, abandoning them without their wages. Before sailing he stole four
patararoes of English ownership, despite the fact he was sailing under

A Dutch armed merchant similar to Van Hoorn's *St. Nicholas*

an English flag at the time. While in Cádiz, Van Hoorn whipped to death an Englishman named Nicholas Browne for no apparent reason.

In the Canaries, Van Hoorn rustled a herd of forty goats. From there, he went on to the Cape Verde Islands, where five more of his men deserted. He then sailed to the Guinea Coast, where he sold some of his guns and powder for gold.

Soon after, he fell in with two of his countrymen, Dutch ships trading in Africa. These he plundered of all they had, a rich take of thirty thousand dollars' worth of booty. In the same area, he stopped an English ship and stole a slave from her, as well as a canoe from Cape Coast, which he plundered, killing three of the black men who crewed her.

With the capital he had raised through indiscriminate plundering, Van Hoorn purchased more than one hundred slaves for export and

sold the rest of his take for a hefty sum of gold. He sailed to the coast
of Capa and involved himself in one of the many tribal wars that
plagued Africa. Van Hoorn's artillery helped his allies prevail in their
fight, and he sailed away with six hundred more captives, presumably
the men and women of the vanquished tribe.

The former sailors from the *Mary and Martha* give a good sample of
Van Hoorn's tactics.

> He did everything under English colors, burning all the houses
> and destroying all the negroes' crops and stores. A month later he
> captured a canoe with twenty negroes, shot one and took the
> rest.[2]

Van Hoorn crossed the Atlantic and called at St. Thomas and
Trinidad, where he sold a number of his slaves. At the end of November 1682, he sailed into Santo Domingo in the present-day Domini-
can Republic. He had approximately three hundred of the blacks he
had brought from Africa; the rest either had been sold or had died
during the horrendous voyage.

Van Hoorn planned to sell what remained of his cargo at Santo
Domingo, but he arrived to find a hornet's nest. The Spanish were
enraged by de Graff's capture of the payroll ship. Van Hoorn's cargo
was confiscated, and Van Hoorn and his crew were detained.

It was some months before Van Hoorn was able to somehow escape
with the *Mary and Martha,* which he now called *St. Nicholas* (patron
saint of sailors and thieves), making his way to Petit Goâve. His only
thought was to make the Spanish pay for having the audacity to take
vengeance on him, and he was looking for men who would join him
in that effort. Petit Goâve was the right place to be.

Petit Goâve was teeming with men who enjoyed nothing more
than plundering the Spanish. It was also a good place to obtain official
sanction for such an enterprise. Van Hoorn found both.

The governor at Petit Goâve was M. de Pouançay, the same man
who had organized the buccaneer contingent that joined Admiral
d'Estrées in his ill-fated expedition to Curaçao. Once again, he rallied
the buccaneers, putting nearly three hundred men aboard Van
Hoorn's ship to aid him in his reprisals.

De Pouançay issued Van Hoorn a privateer's commission, using as
his pretext the complaints of Jamaican governor Sir Thomas Lynch

concerning the pirates and interloping planters at Ile à Vache. Van Hoorn, however, was all set for a cruise of revenge against the Spanish. Apparently not trusting the vicious Dutch captain completely, De Pouançay put the Chevalier de Grammont on board as second in command.

De Grammont had had poor luck in his filibuster career since his spectacular raid on La Guaira in 1680. During the summer of 1682, he commanded a fleet of eight pirate ships which had among its captains Pierre Bot, a Breton pirate who had sailed aboard the ships of the Knights of Malta, and Yankey Willems. For several months, they prowled Cuba's northern shore, hoping to snap up a treasure-laden galleon, but to no avail. They returned to Petit Goâve nearly empty-handed. Fortunately for them, Governor de Pouançay had a job for which their talents were eminently suited.

Van Hoorn, de Grammont, and the rest sailed aboard the *St. Nicholas* from Petit Goâve, in search of others to join their cause, most particularly the renowned Laurens de Graff. They called first at Jamaica to replenish their ship and to deliver letters to Governor Lynch.

Soon after Van Hoorn's departure from Jamaica, Sir Thomas Lynch wrote a fascinating letter that reveals much about the unofficial encouragement of piracy by government officials and the mutual desire of the French and English to covertly harass the Spanish.

The mere fact that the French governor de Pouançay and the English governor Thomas Lynch would happily do business with old reprobates like Van Hoorn and de Grammont says much about the official wink and a nod toward piracy. Lynch, of course, had referred to de Grammont as "an honest old privateer,"[3] a perfect example of how one man's privateer is another man's pirate.

Lynch goes on to say that the *St. Nicholas* "brought me letters from Mons. Poncay [*sic*] and Mons. Grammont . . . [that] assured me of their intentions to keep the peace, and . . . that Van Hoorn, the captain, had no other commission but to take pirates, nor other design here but to deliver his letters."[4]

By "keep the peace," Lynch, of course, meant that Van Hoorn had no intention of attacking English shipping, which was reiterated by his assurance that Van Hoorn had "no . . . other design" in Jamaica but delivering letters and, as he went on, buying medicine and ship's stores.

Lynch is clearly aware that Van Hoorn is not on a peaceful mission. He knows that Van Hoorn's commission from de Pouançay gives the

A Dutch warship of 1680

Dutchman permission only to capture pirates, but Lynch is also certain that Van Hoorn has no such plans. He writes:

> Everyone here concludes that Van Hoorn is also gone to Laurens (the man who, as I wrote to you, took 122,000 pieces of eight off Porto Rico). Van Hoorn has provisions for six months. Nobody thinks he would carry this to capture pirates, nor that he would come to leeward after them when he knows they are to windward.[5]

Lynch is equally aware of Van Hoorn and de Grammont's plans to link up with de Graff and form a powerful buccaneer army. He goes on to say:

"An Honest Old Privateer"

The pirates are all joining Laurens in the Bay of Honduras where he is said to have two great ships, a barque and a sloop of ours and five hundred men. Three days ago I gave the master a letter to Laurens requesting him to punish the pirates and deliver the sloop, which I believe he will do.[6]

Again, Lynch clearly regards robbery on the high seas as piracy only when it affects English shipping. Lynch refers to them all as "pirates," but at the same time he writes a very businesslike letter to de Graff asking that the "pirate" who stole an English sloop ("a sloop of ours") be punished and the sloop returned. He seemed to have had reason to believe de Graff would cooperate with that request.

Lynch's agenda becomes plain when he notes: "For I hope to bring them [the buccaneers] to that pass that they will be content if we do not punish them for robbing the Spaniards. . . ."

Lynch ends the letter with a humorous comment on how much of a Caribbean governor's time was spent dealing with the buccaneers. He writes, "You cannot blame me for being the historian of these rogues for this year, for I have business with few else. . . ."

Sometime in mid–February 1683, Van Hoorn left Jamaica with his small fleet and three hundred buccaneers with the Chevalier de Grammont as his second in command. Governor Lynch's intelligence had been accurate. Instead of beating back to windward and capturing pirates at Ile à Vache, Van Hoorn ran downwind to the Gulf of Honduras to meet with Laurens de Graff, the man who was rapidly becoming the first among equals in the filibuster community.

On the way they met up with others, whom they persuaded to join in on their joint action. Together, their massive buccaneer army would stage one of the most brilliant, if bloody, raids in the history of the Spanish Main.

23

The Documentary That Officially Wasn't

So there we were. All dressed up and no place to film.

Mike Rossiter began a furious series of communications with everyone who might do us some good. He called Antonio, he called the BBC, he called the Venezuelan ambassador whom he had previously contacted.

Communication with the outside was very difficult. We didn't have cell phone coverage. All calls had to be made via radio and ship-to-shore communications, which are primitive and awkward. Mike's job was to get the filming done, and he took every step. Somewhere out there, beyond the sharp line of the horizon, we knew that because of Mike's calls, dozens of people were mobilizing, trying to untangle this bureaucratic Gordian knot.

Ron Hoogesteyn took a more direct, pragmatic approach. The navy ship was some way off and could not see what we were doing on board the *Antares*. Once we were out on the reefs, we would be out of its sight. Ron felt that as long as we were discreet, we might as well start filming.

I was dubious. Could the navy crew really be so oblivious, or so lackadaisical? All they had to do was to come out in their boat and they would see what we were doing. But Ron was a local. As captain of a boat that earned her living in those waters, he knew better than I what to expect from Venezuelan officials, and he knew we were within the law. I figured it was worth a try, but issues of filming were Mike Rossiter's call, not mine.

Mike was torn. As a representative of the BBC, the last thing he wished to do was to embarrass a government agency by ignoring its orders. On the other hand, it was his job to make a documentary. The BBC had already invested quite a lot in getting us out to the site and ready to dive. We had valid permits. No one wanted to see all that money and effort thrown away.

Those things considered, Mike decided we should go for it. I am sure that he would not have made that decision if he had not felt in good conscience that he had done everything required of him to secure the necessary permits. Whatever snafu or bureaucratic meddling had led to the navy's refusal to recognize our permits was not the result of any oversight on Mike's or Antonio's part. While the people in Caracas and London whom Mike had mobilized to straighten this mess out began making calls of their own, banging on doors and cutting through red tape, we prepared to do some diving.

The *Antares* carried a smaller dive boat called the *Aquana,* which was twenty feet long or so. It was capable of carrying a surprising amount of gear. A canopy top provided shade to a small portion of the boat. She was steered from a center console and powered by twin seventy-five-horsepower Yamaha outboards. While their best days were long gone, they could still move us right along with the skiff's flat bottom.

The flat bottom and shallow draft also allowed the boat to get over the reef if the seas were not too high, a great advantage. Still, whenever we approached those treacherous reefs, we had one of the *Antares*'s crew stationed on the bow, warning us of coral heads that even the shallow dive boat would not clear. We did not want to share d'Estrées' fate.

That first morning we loaded our dive equipment and crew aboard the *Aquana,* and then, more discreetly, the video gear. Ron took the wheel, and one of his native crew perched at the bow to keep an eye out for coral. We motored for the reef.

For Chris Macort and me, it was a shock. The last time we had been to Las Aves, the wind had been howling at forty knots and more and the seas had been breaking over the reefs in great showers of foam, flinging spray fifty and sixty feet in the air and preventing us from getting close. This morning, it was as still as a mountain lake. There was no surf breaking, nothing to indicate that the reef was even there. The surface of the ocean was a flat, unbroken plane, from where we sat in the *Aquana* clear to the far horizon. We hardly recognized the place.

This was not entirely a matter of luck. I had spent a lot of time talking to local people about what time of year we would be least likely to encounter that kind of wind again. There is no guidebook or weather forecast that can beat local knowledge, especially the local knowledge of people like fishermen, whose livelihood depends on the weather.

The first time I had been over the reefs, we had anchored inside and fought our way out underwater against the terrible current. Not this time. With no seas breaking on the reefs, and the dive boat's flat bottom and shallow draft, we were able to motor right over them out to the open water. The man in the bow used hand signals to direct Ron around the bigger coral heads, those close enough to the surface that they threatened to rip out the boat's bottom. In that way we threaded our way to open water.

Ever since I first returned from Las Aves I had been studying the charts of the area and overlaying them with d'Estrées' map to try to get a sense of where we might begin to search for the wrecks. The night before I had gone over them once more, making my final decision of where to begin the search.

The ships of d'Estrées' fleet had struck all along the reef, from the southernmost end to the north. In theory we would find wrecks anywhere along the four-mile length. I aimed for midway along the reef, thinking that would put us in the best position to find one. We could then use that as a jumping-off point.

We motored over the shallow water that swirled over the reef. The sea was green when you looked out over the surface, and absolutely clear when you looked straight down. Below us, the mottled blues and browns and yellows of the coral passed slowly under the boat as we headed for the wreck site. And then the reef began to drop away as we passed over to the seaward side and the open ocean.

Without the magnetometer, we had no way to remote-scan for the wrecks. It would be a visual survey. We would pick a spot at random and begin searching north and south until we located the artifacts that would indicate where a ship went up on the reef.

Ron stopped the boat on the seaward side. Before we went in, we took the first logical step in hunting for a shipwreck.

We looked over the side of the boat.

And there it was.

It was that simple. In the clear, shallow water on the reefs of Las Aves, we simply looked down and just below us, on the bottom, fifteen feet or so down, were the ghostly, coral-encrusted shapes of cannons, anchors, the unmistakable signs of a shipwreck. Hidden among the organic shapes of the reef were obviously man-made objects, shapes with right angles, unnatural lumps, things that did not belong. From our seats aboard the *Aquana* we were shouting, "Look! A cannon! There's an anchor!" It was a thrill, all the pleasure of discovery with hardly any of the frustration and setbacks. After years of working the cold, murky waters of Cape Cod, it was a magnificent treat.

Over the side we went to take a closer look. The seas weren't a washing machine this time; we weren't tossed against the unforgiving reef. The water was absolutely calm as we kicked down to the site of the wreck on the seabed.

The wrecks at Las Aves are not what most people think of when they think of shipwrecks. They are not like wrecks in the movies—a ghostly stove-in hull resting on the bottom, broken masts tilted at crazy angles, shreds of rigging draped around. A year or two after they struck the reef the sunken ships might have looked like that, but no more. There is no rigging, no masts, no hulls.

The French men-of-war had been underwater for 320 years by the time we showed up to look for them. That's a long time. The tropical ocean may be a great place to dive, but it is a bad place to be a shipwreck. The very things that make the diving so great—the warm water and abundance of sea life of all kinds—are exactly the same things that quickly destroy whatever is on the bottom.

All of the organic material, the wood hulls and masts, the canvas sails, the hemp and manila rigging, was long gone. What we found were inorganic objects, for example, the stones that had once formed the ballast in the bottom of the ship. These stones tend to be clustered in an arrangement shaped roughly like a football, but narrower. This is how they were when they were piled into the very bottom of the ship. A ballast pile stands out from the rest of the ship bottom because of its shape.

Of the man-made objects, the largest pieces of metal last the longest, and among the biggest artifacts on board these ships were cannons. Cannons were scattered all around. If a ship crumbled straight down, you would expect to see the cannons in a line, as they had once been on the ship's deck. But that is not how these ships fell apart. No doubt some wrecks were lying on their sides and the guns toppled one way or another as the decks rotted away. Most of the three-hundred-plus cannons aboard the fleet were bronze, many about six feet long, but some were bigger, huge guns that you would find on the biggest warships of the day.

Bronze was the preferred metal for guns of that era, but it was enormously expensive. The bronze guns needed to arm a big man-of-war could cost as much as the ship herself, or more. Since they were such special weapons, they tended to be more ornate and beautiful as well.

Bronze remains perfectly intact underwater, even after hundreds of years. The metal alone is quite valuable, but an intact seventeenth-century bronze gun is worth a fortune today, sometimes as much as $100,000. The bronze guns were the things that the treasure hunters

Barry Clifford with anchors;
the hard coral gave them no place to grip

had their eyes on, if gold was not to be found. We found no bronze cannons.

We did find anchors, lots of them. A ship always carried more than one anchor. The largest ships might carry a half-dozen or more, each one of them a monster. The anchors we found were incredible, some with shanks as long as eighteen feet and eleven feet wide from fluke to fluke.

During the course of more than three centuries, these objects have slowly become a part of the living reef. The wreck sites we found each consisted of a long, narrow pile of ballast stones, and nearby, a few anchors, and scattered around the area, some cannons lying at odd angles. That was it.

And that is all that is left of the mighty men-of-war and pirate vessels of d'Estrées' fleet.

24

Where the Wrecks Are

A diver floating in the water above and looking down at one of the wrecks at Las Aves, a diver not accustomed to finding artifacts underwater, would not see much at all. It is very likely that he would not realize he was looking at a wreck. It's like looking at a black cat in a coal bin. The ballast piles are inconspicuous and the cannons and anchors are covered with coral, so that they tend to blend in with the coral reefs, as if they were camouflaged, hiding from would-be salvors.

It takes years of practice to train your eye to see the man-made object among the rest of the debris, but eventually you can see right off when something just does not look right. I can watch thousands of stones tumbling down a sluiceway and I'll pick out a silver coin that's black and concreted. I don't know how. It's as if my mind sees it before my eyes do and my hand will just reach for it. It's the same swimming along a reef. I have spent so much time looking at the ocean floor that when something unnatural is there I can usually spot it. Sometimes it is the shape, sometimes it's the color of the coral, which is different because it is growing on something organic, like a wood hull, or something metal. But mostly it is subconscious.

I think the difficulty of recognizing man-made objects on the bottom became a source of frustration for Charles Brewer. He wanted desperately to see what we were seeing, but he just couldn't. He was out of his element.

If we had been in the jungle, Charles's lair, the situation would have been reversed. Charles would have seen a thousand things that I would never notice. The difference is, I would have said, "Hey, Charles, what's that?" I am happy to be a professional in my own field; I don't feel the need to appear to be an expert at everything. But Charles was different, and his inability to find wrecks chaffed at him.

Disguised as they might have been, the artifacts we saw from the boat were as obvious to me as if they had been labeled. Neither the Venezuelan navy nor the coast guard seemed to take any interest in what we were doing, so we went to work. We had located a wreck site, and there was no reason not to start mapping.

Since we had all worked together on the *Whydah* and other projects, and since we had reviewed our techniques while in Caracas, we had the drill down. In fact, many of the mapping techniques we used at Las Aves had been developed for the *Whydah* site, especially by Todd Murphy.

Todd's background in exercise physiology gave him experience in statistics, and his military training taught him a lot about electronic navigation systems, which we used extensively. What Todd brought to the system was an effort to simplify it, since doing anything underwater is much more difficult than doing it on land. Very often—and this was really true with the *Whydah*—conditions were so bad you simply had to do the best you could as quickly as you could.

Mapping any site begins with what is known as a datum point. This is a fixed reference point, what would be the 0–0 point on a graph. Todd would choose the datum for each site, some point just off to the side of the bulk of the wreck. This is the 0–0 point for local reference—in other words, the starting point for that particular site. Everything we find on that particular site is measured in terms of distance and direction from the datum point. This is the standard grid system right out of "Underwater Archaeology 101."

Once the datum point was chosen, we would physically mark it. If it was in sand we would stick in a pole and that would be our point. If it was on coral we would tie a line to the coral with a buoy attached. As long as there was no current or tide running, the line would stay straight up and down.

The next step was to determine exactly where on the globe the datum was located—in other words, the exact latitude and longitude. Modern technology has made that considerably simpler, particularly the advent of Global Positioning System satellites, or GPS. The system is very common now, but for any who might not be familiar with it, here is a simple explanation.

A GPS receiver on earth picks up signals from multiple satellites in orbit over the earth. From those signals it is able to triangulate the exact position of the receiver, down to centimeter accuracy. When GPS was first available for civilian use in the 1980s, the receivers were bulky, expensive, and unreliable. Now for $300 you can pick up a handheld GPS that is reasonably accurate and about the size of a conventional telephone receiver. We used a highly accurate Trimble Navigation GPS as our primary unit. We also used a British OmniStar GPS.

GPS was developed by the military. Its potential uses are obvious, from navigation on land or sea to running the guidance systems on missiles. Only after the military had fully integrated it to their needs was it released for civilian applications. But the military version was considered too accurate for civilian use. It would have created a serious risk to national security if anyone could walk into a marine supply store and pick up an electronic navigation device accurate

Surveying a wreck site

enough to drop intercontinental ballistic missiles into American missile silos.

To deal with that problem, the government built into civilian GPS "selective availability." Essentially, they threw in an error. For nonmilitary applications, the government would periodically degrade the signal, so that rather than being accurate within a few centimeters it would be accurate to within one hundred feet. That's good enough to find the harbor if you're aboard your yacht, but not good enough for some enemy nation to use as a missile guidance system.

It also wasn't accurate enough for our purposes when it came to geographically locating our datum points. Fortunately, there are ways around the selective availability problem, ways that we had used with *Whydah*. They involve using two GPS receivers rather than one.

Here is how it works. First, you have to know where you are—exactly, down to the fraction of a second in latitude and longitude. This has to be done on land, since a boat moves too much to establish a fixed point.

We pick a station on shore for which we know the precise location, generally by locating it on a chart of the area. Since we know the exact latitude and longitude of the shore station, we feed that information into the GPS at that location. The shore GPS knows exactly where it is. If it then gets a signal collected from the satellites that tells it that its location is, say, twenty meters north of where it knows itself to be, then the GPS knows the selective availability error at that time. It can figure out the error built into the satellite signal.

Out in the boat, right above the datum point, we have what is called a differential GPS, or DGPS. The DGPS is capable of receiving a differential signal—in other words, a signal that will allow it to correct its own reading.

The DGPS over the datum point tells us where it thinks it is, within one hundred meters. But it is also receiving the signal from the shore GPS, and the shore GPS, which knows exactly where it is and what the satellite error is, is saying to the DGPS, in essence, "Hey, the satellite is telling you that you are twenty meters north of where you really are. Correct for that." The DGPS makes those corrections, and the resulting position is accurate not within meters but within centimeters.

That is how we do it at the *Whydah* site, but at Las Aves there was an additional problem. There is only a little landmass associated with the reefs at Las Aves, and we anticipated problems in getting an exact geographical location for the land GPS. Also, we didn't know if we would be allowed to set up a land station. The Venezuelan coast guard station on the island might not want foreign civilians wandering around with sophisticated electronic equipment. For Las Aves we had to work out another way.

The other way was to get the differential signal from a satellite. In an effort to provide extremely accurate GPS data, a number of government organizations, including the U.S. Coast Guard, broadcast differential signals for use with DGPS units. There are also a number of commercial providers. The differential signal is via radio waves, broadcast from radio navigation beacons, commercial FM transmitters, and geostationary satellites.

Along with maritime navigation, one of the primary uses for DGPS is in farming. Farmers use this technology to map their fields and plot where, when, and how they will plant. If you are in the United States or Canada, you can generally pick up one of the government signals for free. Unfortunately, we were using it for a very different purpose, and at a very remote location. We would not have access to one of the U.S. government signals. So we contracted with a private satellite firm for the use of their differential signals. With the satellite sending the correction to the DGPS unit over the datum point, we were able to pinpoint exactly where on earth each of d'Estrees' ships came to rest.

Now, just a few years later, all this effort is unnecessary. With the technology and the ability to correct the signal so easily accessible, the government has removed the selective availability in civilian GPS. Now GPS receivers all have military accuracy, and our work has become that much easier. But we did not have that luxury in 1998.

Ideally, we would have set up a whole network of datum points for each site. In a best-case scenario we would set up a boundary around the whole site and have some datum points inside the boundary and some outside the main circumference. We would map those datum points relative to each other and then start mapping in artifacts relative to those datum points.

In a perfect world, we try to fix each artifact relative to three datum points. That way you can triangulate each artifact and check your measurements, which is much more accurate. But this was not a perfect world, and it only got worse.

With the time constraints and the conditions at Las Aves, we had to go with one datum point and take strikes and measurements from that. We wanted to find as many wrecks as we could, to survey and map them all. We had two weeks. And that was only if the navy did not get serious about enforcing its orders.

We decided to work quick.

25

Unwelcome Intrusions

*I would believe [an attack on Vera Cruz] almost impossible,
except for the experience and valor of those who hear my words.*
—*The Chevalier de Grammont*

SPRING 1683
THE GULF OF HONDURAS

The pirate wrecks we have found, the *Whydah* and others, bear
silent witness to the brutal end that met so many buccaneers. It
was not a career from which many retired peacefully.

Yet not all pirates finished their lives swallowing a lungful of salt
water or dancing at the end of a rope. There were a few who managed
to hit it big and retire, a lucky handful who went ashore with their
fortunes and became wealthy and respected citizens. Laurens de Graff
would become one such man. In the summer of 1682, however, he
still had years of buccaneering left in him, many bloody conflicts, and
many wild, audacious acts.

The capture of the *Princesa* and her 122,000 pesos in Spanish payroll
money was not enough to tempt him to stop. After the capture of that
ship, and the subsequent conversion of the Princesa into his new flagship, de Graff sailed for Cartagena to see what might be found there.
Sailing in company with him was yet another Dutch filibuster,
Michiel Andrieszoon, who would work with de Graff on a number of
pirate ventures.

Cartagena was disappointing. The two Hollanders found nothing but small coasting vessels. The *Princesa,* one might imagine, was a hard act to follow. The buccaneers were looking for far more than what they found off the South American coast. Still in company, they sailed off to the northwest for the Gulf of Honduras, where they had reason to believe that the hunting might be more fruitful.

They were right, to a point. In the Gulf of Honduras they encountered two large Spanish ships, the *Nuestra Señora de Consolación* and *Nuestra Señora de Regla*. These ships, riding at anchor, were part of the regular shipping that moved between Cádiz and the West Indies, bringing wealth and supplies back and forth.

The two vessels had arrived some months before and discharged their cargo, which had been taken overland to Guatemala. Now they were preparing for the return trip to Spain, waiting for the profits from the sale of the cargo to come back from the inland city, along with the goods that they would carry back to the Old World, including valuable indigo and gold.

De Graff, unlike many of the buccaneers, was not a rash or impulsive man. He understood that the ships he found in the Gulf of Honduras were of little value, empty as they were, but in a month or so they would be crammed with treasure and valuable cargo. He and Michiel Andrieszoon left the gulf and sailed to nearby Bonaco Island, where de Graff could careen his ship while their Spanish plum ripened.

Neither de Graff's presence in the Bay of Honduras nor his intentions were much of a secret. Lynch reported that "Laurens . . . lies by to intercept a ship of forty-four guns and four hundred men, with another just half her strength, that are loading goods and money at Guatemala."[1] If a royal governor was privy to this information, it is a good bet that the buccaneers knew it as well. And clearly they did, for Nickolaas Van Hoorn sailed directly from Jamaica to the Gulf of Honduras in search of de Graff.

Instead he found the same two ships de Graff had discovered. Van Hoorn was unable or unwilling to see the sense in waiting until the ships were loaded with cargo. Instead, he attacked.

As it happened, the Spanish were also aware of de Graff's presence and had little of value aboard the ships. It will never be known whether the Spanish would have eventually concluded that de Graff was gone and then sailed into his trap. Long before that could happen, Van Hoorn moved in and captured the empty ships.

Going aboard the larger of the two vessels, Van Hoorn was furious

A forty-six-gun British warship

to find barely thirty chests of indigo in her hold. In a rage, he burned the larger ship and took the smaller as a prize. From there, he sailed off to find de Graff.

De Graff and Andrieszoon were still at anchor at Bonaco Island, waiting for the Spanish ships. During that time, a small privateer unwittingly sailed into the harbor and was taken by Laurens and company, who treated them as prisoners. Some time later Robert Danger-field, a sailor on board that ship, described what happened when Van Hoorn arrived:

[S]eeing two sailes, supposing them to be Spaniards, they [de Graff and his men] gave us our Armes on Condition yt. wee should Waite on Capt. Lawrence and Ingage wth him & undr. his Comand and if they toock a prize wee should have a share

wth. them but Comeing up wth them wee found it was Van
Horne wth his Spanish Prize and soe Lawrence being Disa-
pointed, wee were afraid of being served soe again and soe in the
night left them. . . . [2]

It might be an understatement to say that Laurens was "disappointed." He was so angry that Dangerfield feared that de Graff would
vent his fury on them.

Sir Thomas Lynch, the self-styled historian of pirates, heard the
same story. According to the reports he received, "Van Hoorn . . .
boards the larger of the two ships, finds but thirty chests of indigo,
burns her in a rage, and bringing off the smaller vessel joins Laurens
who was violently enraged at having thus lost his prize."[3]

Van Hoorn was for joining forces—indeed, he had sought out de
Graff for just that purpose—but de Graff was not interested, particularly after Van Hoorn had upstaged him. De Graff had men and ships
enough; he did not need the help of a violent drunk like Van Hoorn.
Again Lynch summed it up, saying, "He [Van Hoorn] has tried to
draw the privateers together, but it is said that Laurens, having two
good ships and four hundred men, will not join him, and that his [Van
Hoorn's] own people and the other French abhor his drunken insolent
humor."[4]

A Gathering of Buccaneers

Despite de Graff's reluctance, the men in his company believed it was
a good idea to join forces with Van Hoorn, and de Graff finally
relented. Perhaps Van Hoorn's second in command, the venerable old
Chevalier de Grammont,[5] played the part of peacemaker, using his
reputation and commanding presence to bring about an accommoda-
tion, grudging though it might be.

The pirates retired to the nearby island of Roatán, there to decide
upon which unhappy Spanish town they would descend. This gathering at Roatán was one of those extraordinary events in pirate history,
like the wreck at Las Aves, five years earlier, which had initiated this
wave of large scale pirate action. On the sandy, jungle-covered island
were gathered more than one thousand buccaneers, among them the
most influential and feared in all the New World.

Here was the Chevalier de Grammont, once the most powerful of the buccaneers, now relegated to vice-admiral status. Here was the mulatto Laurens de Graff, whose very name filled people of the West Indies with terror, and would continue to do so for decades. Here were the vicious killers Nikolaas Van Hoorn, Yankey Willems, Michiel Andrieszoon, Pierre Bot, and Jean Foccard.[6] It was one of the largest gatherings of buccaneers ever, a prime example of the fluid alliance that existed among the Brethren of the Coast.

The Gulf of Honduras offered little opportunity for the buccaneers. Their presence was well known, and every city and naval vessel in the area was on the alert. With reinforcements on their way from Cartagena, they knew they had to leave the area and fall on some unsuspecting city before word of this massive gathering spread.

They decided on Vera Cruz.

26

The Sack of Vera Cruz

[I]t is not right to behead any surrendered man who has been granted quarter.
—Laurens de Graff to Nikolaas Van Hoorn

MAY 17, 1683
VERA CRUZ

It was not an easy decision to reach. Vera Cruz was a well-fortified city, nearly as strong as Havana or Cartagena. It had not been attempted since the Elizabethan sea dog John Hawkins, mentor to Francis Drake, had staged an impromptu raid in 1568. That had gone badly for the attackers. With de Grammont's eloquent assurance that no Spanish force could resist their onslaught, however, the buccaneers agreed.

Vera Cruz held a great deal of potential, along with possible danger. It was to that city that much of the wealth of Mexico and Central America was shipped, before being sent to Spain. Every year a fleet of massive galleons, known as the plate fleet, arrived to transport the accumulated wealth across the Atlantic. With luck, the pirates would hit while the treasure still lay in the storehouses, and before the great men-of-war arrived to carry it back to Spain.

About the same time that their old compatriot Thomas Paine was leading his attack on St. Augustine, the pirates sailed en masse from Roatán Island, making their way due north to weather the Yucatán Peninsula. On April 7, 1683, they went ashore at Cabo Catoche, the northernmost point of the Yucatán, to make their final arrangements before descending on Vera Cruz.

Command of the venture was never firmly set. The ultimate authority seemed to rest with Laurens de Graff, though some accounts list Van Hoorn as "General"[1] in command of the main body of men. It nonetheless became clear that de Graff was calling the shots.

This was only reasonable. No one, not even his own men, could stomach Van Hoorn. Almost two months earlier, Lynch had written that "the French abhor him [Van Hoorn] for his insolence and passion, and they . . . will desert him at the first land or make Grammont captain. . . ."[2]

Lynch was wrong only in thinking that the Chevalier was the heir apparent to the leadership of the buccaneers. He was not. It would soon become clear that de Graff had ascended to that lofty place. The torch had been passed.

Soon the pirate armada was under way again. Among the vessels in the fleet were two ships captured from the Spanish, the *Nuestra Señora de Regla*—the ship that Van Hoorn had whisked from de Graff's trap—and a prize taken by Yankey Willems. These ships were filled with a large contingent of buccaneers, certainly no fewer than two hundred, with some accounts setting the number as high as eight hundred. With de Graff in command of the *Regla* and Yankey in command of his prize, the two former Spanish ships took the lead, leaving the rest of the fleet just below the horizon.

In the late afternoon of May 17, the two erstwhile Spanish vessels appeared off the harbor mouth of Vera Cruz. To the buccaneers' relief, there was no sign of the plate fleet. The twelve large, heavily armed men-of-war were due at any time on their annual voyage to fetch the precious metals of the New World and to carry them back to Spain. They would have been a formidable enemy, but fortunately for the pirates, they had not yet arrived.

Just as fortunate for the pirates, the lookouts in the port of Vera Cruz thought the strange vessels were part of the plate fleet, which was afraid to make their way into the harbor in the failing light.[3] Rather than sending a vessel to confirm that this was in fact the case, the lookouts lit fires on shore to help guide the ships safely in. De Graff made good use of the Spaniards' courtesy and stood in the harbor, anchoring near shore. His disguise had worked, and the wolves were in among the sheep.

In the early hours of May 18, de Graff and Yankey slipped ashore with the large force of buccaneers they had aboard their two ships. They silently reconnoitered the town, trying to get a feel for its defenses. Vera Cruz was a city of around six thousand inhabitants. Of those, four hundred were civilian militia and another three hundred regular troops, with three hundred more garrisoned on the island fort of San Juan de Ulúa. The pirates were evenly matched. If the Spaniards mounted any sort of decent defense, it would be a hard fight.

De Graff and Yankey's force, as the "forlorn," also known as the "forlorn hope," was to be the first over the wall, the first through the breach, while Van Hoorn and de Grammont landed their men some distance away and marched in support. The job of "forlorn hope" was potentially as bad as it sounded.

On the landward side of the town stood two forts that were de Graff's target. There sand dunes had drifted up against the stockade fences, making it a simple matter for the pirates to slip over that first line of defense. Once in the forts, the pirates encountered the usual degree of alertness among the Spanish troops and sentinels: they were all asleep.

During the night, Van Hoorn and his forces joined up with de Graff, and at dawn they attacked. The pirates fired wildly and indiscriminately and set the entire city in a panic. They kicked in doors, fired at anyone who showed his face, cut down any armed men who appeared. The soldiers and militia fled. After half an hour, the buccaneers held Vera Cruz. They had lost only four men, three of them de Graff's men who had been accidentally shot by Van Hoorn's contingent.

Fearing the possibility of a Spanish counteroffensive, de Graff and de Grammont saw to organizing a defense. In a move that was perhaps a throwback to his "chevalier," or knightly, heritage, de Grammont organized a buccaneer cavalry using horses from the stables of Vera Cruz.

The buccaneers herded as many people as they could—several thousand—into the cathedral and held them prisoner there for three days, with little food or water, while they set about plundering the city. Packed in, with barely room to sit, many prisoners perished, particularly children, as the pirates ransacked the town.

The take was disappointing, and the filibusters reckoned that there was more to be had, hidden in the countryside.

On their second day of sacking Vera Cruz, the plate fleet appeared on the horizon, tipping the balance of force to the Spanish. Also that morning, a line of Spanish irregular cavalry appeared at the western end of the city. De Grammont charged with his mounted buccaneers, flags waving and trumpets blowing. The Spanish were so startled by this unorthodox and bizarre attack that they scattered without a fight.

With the appearance of the Spanish cavalry and the plate fleet, the buccaneers knew that time was running out. Fast and hard "persuasion" would have to be used to locate the hidden wealth of the hold-outs.

The raiders turned their attention back to their prisoners in the cathedral. They selected any prosperous-looking citizens, and their servants, and dragged them from the crowd of prisoners. One by one, they began their systematic torture, using well-tested methods to extract the location of hidden treasure. The pirates threatened to burn down the cathedral, prisoners and all, if more loot and ransom was not forthcoming.

The people of Vera Cruz had no doubt that they were serious in their threat. As one writer put it:

[T]hough at this time [by the third day] they got abundance of Jewels, Plate, etc. and about three hundred and fifty Bags of Cochenelle,[4] each containing one hundred and fifty or two hundred pound weight, as they say; yet were they not satisfied, but put the considerable people to ransom, and threatened to burn the Cathedral and Prisoners in it, which were five thousand and seven hundred, if they did not immediately discover all they had; so that the fourth day they got more than the other three. . . .[5]

The pirates also garnered an additional seventy thousand pieces of eight for the ransom of Governor Don Luis de Córdoba. De Córdoba was discovered hiding in a pile of hay in a stable by a pirate captain named George Spurre, one of the few English captains in the pirate fleet and a man who had been an active buccaneer for about ten years. It was only on Spurre's urging that the governor was ransomed at all and not killed outright by several of the French buccaneers who had once been held prisoner there. As it happened, Spurre's protection only bought de Córdoba a little time. Soon after, he was sentenced to beheading by his own government for allowing the city to be so easily taken by pirates.

By the fourth day of their sack of Vera Cruz, the buccaneers knew

Extorting tribute from the citizens

it was time to go. The powerful plate fleet was slowly bearing down on the harbor. There was also reason to suspect that reinforcements were on their way from Los Angeles, a city ninety miles away.

Having not received all of the ransom due them, the buccaneers marched their prisoners, whom they made carry their loot, and fifteen hundred blacks and mulattos back on board their ships. They sailed from the town of Vera Cruz to a nearby island, there to await the rest of the ransom. The island, home of an ancient Aztec temple, was called Los Sacrificios, giving one a good idea of what could happen there.

The pirates waited nearly a week for the ransoms to be delivered. Van Hoorn, growing impatient with the delay, decided to send ashore a dozen of their prisoners' heads as incentive for the Spaniards to expedite matters.

De Graff, who had the reputation of being more humane than Van

Hoorn—more humane, in fact, than most filibusters—would not allow this. He and Van Hoorn quarreled, and Van Hoorn pulled his sword. The two buccaneers went for each other with cold steel.

De Graff drew first blood, a slash across Van Hoorn's wrist that was not serious, but did put an end to the dispute. The prisoners kept their heads.

At last, having wrung from Vera Cruz all that they were likely to get, the pirates loaded their plunder, hostages, and slaves, weighed anchor, and set sail, with only the plate fleet of twelve heavy men-of-war between themselves and freedom.

Though Van Hoorn raved and yammered about attacking the plate fleet, de Graff refused outright, and the men concurred. Braced to fight their way out of the harbor, the pirate flotilla set sail. To their surprise, the Spanish admiral, Diego Fernández de Zaldívar, did not engage them, but rather let them sail right on by. Why he failed to attack is not known. Sir Thomas Lynch would later ironically suggest that the admiral and vice admiral who had commanded the plate fleet "deserve to be made grandees for allowing these pirates to escape when they had them in a net."[6]

The buccaneers left Vera Cruz with a fortune in loot, with only four of their own men dead in exchange.

So the treasure was divided

Two weeks later, the Vera Cruz raid claimed its fifth and final casualty. The wound that Van Hoorn had received in his duel with de Graff was barely a scratch. But the scratch turned gangrenous and the infection spread. About fifteen days after sailing safely out of Vera Cruz harbor, Van Hoorn died, leaving his son his share of the take, an estimated twenty thousand pounds sterling. A year later, Governor Lynch reported that Van Hoorn's son, too, had died, at Petit Goâve, and the French buccaneers had divided his inheritance among themselves.

On June 24, 1683, the Dutch pirate's men rowed his body ashore at Isla Mujeres and buried him in an unmarked grave, bringing to a close the short and vicious piratical career of Nikolaas Van Hoorn.

27

The Search Continues

We had another official visit our second day at Las Aves, but this time it was the coast guard. I imagine that there was some interdepartmental rivalry going on, that the coast guard would not allow the navy to do all the passport and permit checking, especially not at a place that was home to a coast guard station. It might have been simple curiosity as well.

The coasties came out in the big open boat that they used around the island. They threw us a line, and we tied their boat up. They climbed up to the afterdeck of the *Antares,* about ten in all. They were young men; I doubt that the officer in charge had seen thirty. We invited them into the salon.

The coastguardsmen were pleasant, even apologetic. They tried to be official, but the effect was lost since their "uniforms" consisted of brown pants, white undershirts, and ball caps. If there was interservice rivalry, they definitely lost to the navy in the uniform department.

Uniform or not, they did have authority. Once again, we produced our papers, passports, film and expedition permits—all of the paperwork we had. The coastguardsmen took it all and examined it carefully, then reiterated the navy's position that we could not film.

Charles tried his best with them. We also had several lawyers on board, including Max's friend Pedro Mezquita, and they jumped into the fray, too. The coasties' position was as intractable as the navy's. No filming. We thanked them, shook hands, escorted them back to their boat, then got ready for another day of shooting.

Mike Rossiter maintained his barrage of phone calls. People in London and Caracas were telling him that it would be straightened out, whatever problem had arisen would be solved, and that we would get our permits recognized. We only had food and water and a budget for two weeks. We could not sit on our hands waiting for the bureaucratic mess to untangle. We suited up and loaded our gear onto the *Aquana*.

Our goal was to find and map a wreck a day. It was a brutally demanding plan, so tiring that Carl compared it to basic training with the SEALs. It would not have been possible if our team had not been so experienced and used to working together. During the planning stage of the expedition, Charles had wanted to hire local divers. While there are plenty of very good divers in Venezuela—Ron Hoogesteyn, for example—it would have been impossible to finish as much work as we did without a team that had experience and time together. As it was, we were able to maintain that pace for nearly the entire expedition.

I suspect that Charles wanted to staff the project with people who answered to him, not me.

Before leaving Provincetown, Todd Murphy and I had determined how we were going to run the operation, what equipment and techniques we would use. But you can't plan everything until you see the site. Although I had been there before, the wind and seas had prevented me from the kind of reconnaissance that would have allowed for more meticulous planning. We knew what we wanted to accomplish, and we determined how we were going to execute the operation once we were there.

Once at Las Aves, Todd and I would get together in the evening and work out the plan for the next day. Todd, as director of operations, would then plan who would be diving, what equipment they would need, when we would leave.

That freed me to set the sequence we would follow in exploring the reef, to study d'Estrées' map, and generally to keep the expedition on track toward achieving the goals we had laid out.

Todd and I were both doing the kind of work we love. Todd, in particular, loves the logistics, the planning and the coordination and the teamwork. I'm not sure if he feels that way because he is in Special Forces, or if he is in Special Forces because he feels that way. The end result is the same. He loves his work and he is good at it.

I enjoy the logistical side of expeditions as well. There is something about planning for a trip that whets one's appetite for the journey itself and prepares one for the rigors to come. I imagine Columbus might well have been the same way when he was preparing to set out for Cathay. And, if you are at all attuned to problem solving as I am, I can recommend no better exercise than working up the details of an expedition. More than logistics, I love hands-on exploration; being in the water with the wrecks and seeing them for the first time.

Different explorers have different interests. Some, like Bob Ballard of *Titanic* fame, specialize in deep-water, heavily mechanized exploration. He does his work with submarines and remote operated vehicles (ROVs), and he is good at it. But it is exploration performed from a control room, looking at a video monitor, entirely apart from what is being explored. It's like being an astronaut who is allowed to orbit the moon but not land.

Ballard has made some impressive finds without ever getting wet. That kind of exploration is just not my cup of tea. I enjoy the physical aspects of diving, being my own, human-powered vehicle. The *feel* of a site is as important to me as its *look*.

People have often asked me what is the deepest I've ever dived. The answer is not very. There seems to be a mystique about deep diving, but I explain—half tongue in cheek—that going deep consists of nothing more than strapping on a weight belt and sinking. No special skill involved in *that*.

My real objection is that bottom time on a deep dive is severely limited, making it very impractical for archaeological work. Deep diving for its own sake strikes me as a macho thing and no more. There is an element of danger and unpredictability to it that fall outside of what I consider reasonable and prudent boundaries—especially since the physiological changes that occur at such depths are not fully understood by medicine.

For me the real thrill is the hands-on exploration; going down through the water to see what is there. Fighting the currents to get over the reef at Las Aves was a job for a human being. A machine could not do that. We were powering ourselves, hauling our own equipment. We pride ourselves on our swimming ability, on being our own ROVs. I insist on all our expedition team members being in top physical condition, which is not just a matter of conditioning the physique but also a matter of conditioning the mind-set.

Our daily routine was simple. In the morning we would have a breakfast meeting to inform the rest of the crew about the plan for that day. Since we had all worked together so long, and since we were doing essentially the same work at every wreck site, we were very informal. Sometimes we would issue a "frago," short for fragmentary order, which is a change to an already issued order. The needs of the BBC crew obviously often played a part in our decisions regarding the work of the day.

With the workplan set, we loaded up the *Aquana* and motored over the reef to the dive site. Each member of the team had a specific job, and we kept with those assignments for the duration of the expedition.

We knew the wrecks were strewn along the reef, so we would form teams of two or three divers and swim in tandem along the edge of the reef, searching for artifacts. Ideally we would find the point of impact, the place where the ship first hit. Knowing the size of the vessels, we looked for that point in about fifteen to twenty feet of water. If we found where the ship hit first, we could then look for the scatter pattern of the artifacts. That would tell us a great deal about how violent the impact was and how the ship broke up.

Unfortunately, after more than three hundred years, things have become obscured. More often we would find only the usual signs: ballast piles, cannons, anchors, sometimes parts of the rudder assembly, barely discernible among the coral.

Debris fields were everywhere. Some looked as if the ships had broken apart. Some looked as if they had sunk fairly intact and rotted in place. There was only one thing that was consistent: the wrecks all showed that the magnitude of the disaster at Las Aves was stunning.

28

Mapping

The first thing we did at each wreck site was to establish a datum point. Then we would determine the geographic location of the datum using the DGPS. As soon as the datum was set, the measuring would begin.

Todd Murphy and Carl Tiska did the bulk of this work. They made a great team despite (or perhaps because of) the inherent rivalry between the Green Berets and the SEALs. Using a one-hundred-foot tape measure, they measured from the datum point to each of the artifacts they could observe. Since we had the Agas and the com system, they could relay this information immediately to the field archaeologist, Cathrine Harker.

Cathrine Harker was aboard the *Aquana,* which, if the seas permitted, was anchored as close to the wreck site as possible. Cathrine has fair skin, and the tropical sun was dangerous. She wisely dressed herself in oversized white men's dress shirts, with turned-up collars and long sleeves. These, a floppy hat, and industrial-strength sunblock kept her protected while she worked all day under the brutal rays.

Along with bringing her aboard the *Whydah* project and the Las Aves expedition, I was responsible for another big change in Cathrine's life. Not long after she came to help out on *Whydah,* I told

Carl Tiska and Todd Murphy surveying a wreck site

myself, "Cathrine, I'm going to introduce you to the man you are going to marry."

Then I took Chris Macort aside. "Chris, I am going to introduce you to the woman you are going to marry." In 2000, Chris and Cathrine were married in a castle in Scotland.

Once the datum point was established, Todd and Carl would begin to systematically measure the site. They would measure from the datum to the cascabel, the round knob at the end of each cannon, and from the north end of the ballast pile to the south end. They would measure from the datum to each anchor. They would communicate every measurement to Cathrine through the single-sideband radio mics in their Agas. Their conversation would sound something like this:

"From the datum to anchor number one is fifteen and a half meters. . . ."

Carl would roll up the long tape measure, then he and Todd would start in on the anchor itself.

"Cathrine . . . the width of the fluke of the anchor at the widest part is . . . one point eight inches. . . ."

"Cathrine . . . the fluke on the second anchor is . . . one foot, nine inches. . . ."

"Circumference of the shaft is . . . one foot, seven inches. . . ."

As Todd and Carl measured each artifact, they would give the artifacts numbers and tag them. Then Chris Macort, working with waterproof paper and marker and a compass, would make a crude drawing of each of the objects, noting the orientation in which it lay.

Chris was originally designated as safety diver for Eric Scharmer, the underwater videographer, meaning that he would work with Eric and make sure Eric didn't have any problems. Chris has a strong artistic streak, however, and that, combined with his experience in archaeological diving, not to mention his unusually close working relationship with the archaeologist, made him the perfect choice for the rough mapping work.

Chris would swim from artifact to artifact and draw stick-figure cannons and anchors, making sure to get their orientation exact. He would do the same with the ballast piles and anything else on the site, like parts of a rudder or clusters of cannon balls and musket balls. When he was done, they would remove all of the numbered tags.

We measured as accurately as we could under the conditions, but there were inherent problems. For example, each cannon was measured from the cascabel to the muzzle. But the cannons were heavily encrusted with coral, built up over the centuries. Aiming at "zero site trauma," we did not want to disturb the coral, so we had to estimate where, under all that growth, the cascabel began and the muzzle ended.

We did the best we could, and as much as we could within the mission parameters. We did three trips a day to the reef. After breakfast we would go out and work all morning. We would come back for lunch and fresh tanks, then go out again. When those tanks ran out, we would return for more fresh ones and make our last trip of the day.

Max and his friends, including Paul Ryan, Kent Correll and Pedro Mezquita, also proved helpful. Max's energy and enthusiasm were infectious. He was having a lot of fun, and I was glad he was able to do so. I would not have heard about Las Aves, and certainly would not have been there looking for wrecks, had it not been for Max. But he and his friends went further. They assisted Todd and Carl in some of the measurement work, and Paul Ryan did a lot of underwater photography.

We never spent more than one day on any one wreck. By the end of the first week we had thoroughly mapped five sites.

In the evenings, while Todd and I discussed the targets for the next day, Chris and Cathrine made revised maps from the data she had collected and the crude drawings that Chris had made underwater. By setting the datum point on the center of a paper grid they could use the measurements to position the drawings of cannons, anchors, and other large artifacts on the map. Chris's notes gave them the exact orientation and size. With the data collected during the day, they were able to produce an exact map of the wreck site every evening. You could look at their finished product and say, "Yes, that is exactly what it looked like." These maps were among the most valuable research that we took from Las Aves.

We started about midpoint on the reef and worked our way north. We would swim along with snorkels to conserve air. Looking down through the shallow water we could see the wrecks from the surface. As the reef fell off toward deeper water, we had to go down on air to find them.

When we found the first wreck, I compared its location to d'Estrées' map. There was a wreck marked on the map very close to where we had found the artifacts. Interesting, but not enough to prove anything.

The second wreck we found, on the second day, also had a corollary on the French admiral's map. By the third day and the third wreck, I was starting to feel confident that my hypothesis was correct: d'Estrées' map was not just a burst of artistic fantasy but in fact a very accurate reproduction of the reefs and the positions where his ships met their ends.

As time progressed, we saw how that theory was really coming together. Every time we found a wreck, we could see it on d'Estrées' map, right where it was supposed to be. *Bourbon, Bellinguer, Defenseur,* and the rest—there were wrecks at the site where each was marked on the map. By the end of the expedition, we were no longer using the wrecks to determine the accuracy of the map but rather using the map to determine where to look for wrecks.

That was important. It meant that d'Estrées' map could be considered an accurate and reliable primary source document when describing the history of the wrecks at Las Aves. It also meant that we could be fairly certain that the wrecks named on the map were the wrecks we were finding. In other words, when looking at a scattering of guns and anchors, we could identify which ship they had once belonged to.

That was most important to me because of the two wrecks identified only as *flibustier*, the pirate ship wrecks. Now that the good admiral had proved himself a reliable map maker, it was a different story. If I found wrecks noted as *flibustier* exactly as d'Estrées had marked them on the maps, I could be reasonably sure I had found more pirate shipwrecks.

I was waiting for my chance to go and look.

29

War and Peace

While you behave with such respect to the justice and friendship
that exist between the French and the English crowns, I am
always your friend.
—Sir Thomas Lynch to Laurens de Graff

Ironically, the pirates more than three centuries ago seemed to get
more government cooperation than we did. And all we wanted to
take was pictures.

The unofficial sanction that the buccaneers enjoyed, however, was
going to come to an end, at least temporarily. The guns of Vera Cruz
reverberated all around the Caribbean, and echoed through the cen-
ters of government in England, France, Spain, and Holland.

In 1678, the Treaty of Nijmegen had been signed in the Netherlands, bringing to an end the third of the seventeenth-century Dutch
Wars, the conflict in which d'Estrées had lost his fleet on Las Aves and
marking the beginning of the great wave of piracy that followed in the
wake of that disaster.

During the five years of peace, five years of suspicious, uneasy, brittle peace in Europe, the shifting alliances and balances of power left
everyone waiting for the next, inevitable conflict. Far from the eyes
and control of their home governments, the power brokers of the
Caribbean continued to play their clandestine games.

Spain Fights Back

By the time of the Vera Cruz raid, Spain was an old, worn-out lion, lacking the strength and skill of its youth. Insulted repeatedly, it could do little more than growl and swat at its tormentors. But Vera Cruz was one insult too many, and Spain fought back.

Spanish efforts at retaliation met with mixed results. The Armada de Barlovento did manage to capture two of the lesser pirate captains who had been at the sack of Vera Cruz. Pierre d'Orange and Antoine Bernard were French filibusters who commanded, respectively, the *Dauphin* and the *Prophète Daniel,* tiny pirate ships mounting two guns each. When they heard of the great buccaneer gathering at Roatán they abandoned their plans to go turtle hunting and instead joined in with Van Hoorn and de Grammont.

The Armada de Barlovento captured the two pirates at Little Cayman Island on August 4, 1683. Spanish law dictated that pirate leaders be tried at the scene of their crimes. The two men were returned to Vera Cruz, where they were confronted by plenty of witnesses, especially d'Orange, who had been responsible for keeping the prisoners locked in the cathedral, where so many had died. D'Orange was asked at his trial how a Catholic such as he could have looted and defiled a cathedral and participated in such horrendous crimes. His answer, to the effect that "everyone else was doing it," did little for his defense.

D'Orange was found guilty. Presumably Bernard was as well, though there is no record of what became of him. On November 22, 1683, d'Orange was marched through the streets of Vera Cruz and, by way of example to other pirates, hanged, then decapitated. His head was put on a spike at the wharf.

Understanding the potential diplomatic ramifications of the sack of Vera Cruz, Sir Thomas Lynch wrote to the governor of Havana, protesting his innocence regarding anything having to do with the event. He explained that he had, in fact, attempted to warn the Spanish of the impending attack. It was only Spanish bungling, he claimed, that had caused the warning not to arrive. The letter is a masterpiece of artistic smoke-blowing. Lynch writes:

One of our men-of-war at St. Domingo demanded Vanhorn as a rebel and a pirate, to bring him here, where he would have received the reward due to such ruffians; but the President [Pouancay] thought fit to protect him, and afterwards released him, having taken 20,000 pieces of eight from him on pretense of the six patararoes he took in Spain.[1]

It seems odd that Lynch would go to the trouble of sending a man-of-war all the way to Santo Domingo to arrest Van Hoorn when six months earlier Van Hoorn had come calling at Jamaica with letters from de Pouançay to Lynch. Lynch had been perfectly aware of Van Hoorn's intentions. Nonetheless, Lynch contends, "I have at my own charge chased out of the Indies all the pirates that prey on us or on your nation. I have done all in my power to serve the Spanish nation."[2] This from the man who six months earlier wrote that he hoped to placate the pirates by not punishing them for robbing the Spaniards!

Lynch goes on to bemoan the fact that for all the help he has given the Spanish, "I have received neither thanks nor civility, nor have the English received any privilege. Not one of our ships, that the Spaniards meet with, will they fail to take and plunder if they can."[3]

Unimpressed, Charles II of Spain declared war on France, in part because of French intrusions in the Spanish Netherlands and partially for the outrage of Vera Cruz, which was carried out under French commissions issued by de Pouançay.

Spain was in no position for a prolonged war unless she was joined by the other nations of Europe, and those nations, not having a dog in that particular fight, declined. After six months of hostilities, Charles II was forced to sign the Truce of Regensburg, ending the conflict.

Despite being the dominant power in Europe, Louis XIV was not interested in war, at least not during the years 1683–84, and he took action to appease the Spanish. In the West Indies, Louis had always maintained an official policy of refusing commission to filibusters, while at the same time cheerfully allowing de Pouançay to issue them under the table. On one hand, he could claim that he was not sponsoring such mischief, while on the other he could use the buccaneers to keep Spain off balance and transfer Spanish wealth into the economy of France.

After Vera Cruz, however, Louis tried to appease Spain. In the Caribbean, this meant dropping the "wink and a nod" policy toward the buccaneers and dampening de Pouançay's cheerful issuance of commissions that rendered their attacks quasi-legal.

De Pouançay died in 1683, making it unnecessary to recall the governor. His successor, Sieur de Franquesnay, reversed the laissez-faire

A Dutch galliot

approach toward the buccaneers and made a genuine effort to implement the Versailles policies aimed at suppression of piracy. It did not go over well.

Some of the pirates simply abandoned Petit Goâve and moved operations to Jamaica, domain of the amiable Sir Thomas Lynch. Thomas Paine, tired of the filibuster's life, had returned to Rhode Island. Many others abandoned the Caribbean completely, crossing the Isthmus of Panama and plundering the Spanish on the Pacific side, easing enforcement problems for France and doubling them for Spain.

A potentially greater hazard for France were those buccaneers who began to look to England for the kind of unofficial sponsorship and succor they had received from France. Among them was Laurens de Graff, the most dangerous man in the Caribbean. In the fall of 1683 de Graff wrote to Thomas Lynch from the now inhospitable shores of Petit Goâve, in answer to a communiqué that has since been lost:

I am much obliged for your civility, and thank you for the honor which you have been pleased to do without any merit of my own. I beg you to believe me the most humble of your servants, and to employ me if there be any place or occasion in which I can be of service to you. You will see how I shall try to employ myself. If by chance I should go to your coast in quest of necessities for myself or my ship, I beg that my interests may be protected and no wrong done me, as I might do so if the opportunity presented itself for doing you service. Begging you to do me this favor, I remain your most humble and affectionate servant.[4]

De Graff would have the opportunity the following year to render his favor to Sir Thomas Lynch. In the interim, the restless freebooter could not stay idle. Though de Graff's pockets were presumably still full of the booty of Vera Cruz only a few months past, the energetic pirate was soon back in business. While Sir Thomas Lynch and the Lords of Trade and Plantations were putting together an offer to entice the pirate into their camp, de Graff laid a course for Cartagena.

THE BLOCKADE OF CARTAGENA

Laurens de Graff was by now one of only a handful of pirates in Caribbean history who commanded not a ship but a fleet. There were seven vessels in his squadron when he sailed for present-day Colombia. In those ships were many of the men who had already spent years with Laurens, including fellow Dutchmen Yankey Willems and Michiel Andrieszoon.

In late December 1683, the buccaneer fleet arrived off the harbor of Cartagena. Wisely, the governor of that city, Juan de Pando Estrada, decided to stop the pirates before they landed, rather than count on the city's defenders to hold off a land assault. After all, the Spanish track record in such actions was none too encouraging.

Estrada commandeered three private ships for the job. The largest, the *San Francisco*, was a ship of forty guns. The second was called *Paz* and mounted thirty-four great guns. The last, a somewhat smaller vessel called a galliot, mounted twenty-eight guns. Aboard these ships, the governor put eight hundred soldiers. All in all, it was a formidable force.

The ensuing battle was a terrible, bloody farce, made worse, no

A naval action

doubt, by the fact that the Spanish squadron was under the command of twenty-six-year-old Andrés de Pez y Malzárraga, who had only been promoted to captain the previous summer.

The three large Spanish ships, clumsy to begin with and most likely hampered by the great crowds of men on deck, were completely overrun by the smaller, more nimble pirate vessels.

The *San Francisco* soon ran aground, rendering her defenseless against ships that could lie in a place where her guns would not bear and pound her to kindling.

The *Paz* fought for four hours—a noble effort, when one considers that many of the greatest ship-to-ship actions were over in less than thirty minutes—but at last she too struck. Yankey Willems took the galliot. The Spanish lost ninety men killed, the pirates twenty.

De Graff refloated the *San Francisco* and gleefully took her over as his new flagship, renaming her *Fortune,* which was later changed to *Neptune*. Michiel Andrieszoon was given command of the *Paz,* which he renamed *Mutine*. Willems took command of de Graff's former flagship, the former *Princesa*. With the addition of three powerful vessels, the pirate fleet was now ten ships strong.

On Christmas Day 1683, de Graff set Captain Pez y Malzárraga and the other Spanish prisoners ashore. They carried with them a note from de Graff to the governor, thanking Estrada for the Christmas gifts.

Rather than attack the city de Graff decided to blockade the port, hoping to snatch up a valuable prize trying to get in or out. He probably realized that with the element of surprise entirely lost, the people of Cartagena would have long since carried themselves and their valuables far inland, leaving little behind worth taking. Reinforcements would also be on their way to augment the large contingent of soldiers already there.

In mid-January 1684, a small convoy did arrive at Cartagena. It was an English convoy, however, a small fleet of slavers escorted by the man-of-war HMS *Ruby*. England and Spain were at peace, as were England and France. For that matter, France (for whom de Graff ostensibly fought) and Spain were still at peace, as far as de Graff knew. News of Spain's latest declaration of war with France had not reached the filibuster.

Even if England and France had not been at peace, de Graff, as we have seen, was not interested in attacking English shipping. He was only interested in plundering Spain, the country he loathed, and war or peace made little difference to him.

De Graff did not meddle with the English convoy, except to have the officers aboard his ship as dinner guests. As it happened, among the visitors was a trader carrying a letter to de Graff from his wife, Petronila, in the Canary Islands. Through his wife, the Spanish authorities,

Pirates boarding a Spanish ship

eager as the English and French to obtain the loyal service of the great buccaneer, offered him a pardon for all his piracies if he joined the forces of the king of Spain. A former captive of Spain, public enemy number one, he was now being offered not only a pardon but, in effect, a commission as an officer in the Spanish navy.

De Graff had not seen his wife for many years. No matter how much he loved her, he must have realized that he would never see her again, not as long as she was in Spanish territory where he was a wanted man. Now there was a chance to see her again, to free himself from the threat of Spanish reprisal and Spain's relentless pursuit of him. In the end, however, de Graff simply did not trust Spanish promises. He made no response to the offer. He perhaps was wise.

Soon after, the pirates gave up the blockade and headed north. While under way, de Graff came upon two vessels, which he followed from a distance until nightfall. Under cover of dark, he fell on one of the ships, boarded her, and took her with only two shots fired. She turned out to be a Spanish vessel of fourteen guns, carrying quinine and nearly fifty pounds of gold. Laurens de Graff was back in business.

The next morning de Graff took the second ship. This one turned out to be English, laden with sugar, which the Spanish ship had illegally captured and was escorting to Cuba. De Graff had his opportunity to render the English a service, and this he did, by releasing the former crew from their captivity and setting them and his prize free.

The gesture did not go unnoticed. Lynch had already reported the blockage of Cartagena to the Lords of Trade and Plantations and speculated that the buccaneers might use their now expanded fleet to raid Vera Cruz again.[5] His letter reflects thinly disguised animosity toward the Spanish, and an ambiguous sense of duty.

30

The Wooing of
Laurens de Graff

But Scripture saith, an ending to all fine things must be;
So the King's ships sailed on Avès, and quite put down were we.
All day we fought like bulldogs, but they burst the booms at night;
And I fled in a piragua, sore wounded, from the fight.
—"THE LAST BUCCANEER"
Charles Kingsley

SPRING 1684
PETIT GOÂVE, HISPANIOLA

Pirates continued to be a mixed blessing for the governor. Lynch was happy to entice French pirates like de Graff and the "honest old privateer" de Grammont, but he chafed at the activities of English pirates and the support that they received in the North American colonies.

From Charlestown, Massachusetts, to Charleston, South Carolina, pirates were very welcome, and would continue to be. The Navigation Acts prohibited the colonists from trading with almost anyone except the mother country. The cash-and-merchandise-strapped colonists naturally welcomed pirate booty with open arms and purses. Lynch griped:

I have formerly advised you that our laws against privateers neither discourage nor lessen them while they have such retreats as Carolina, New England and other colonies. They have permitted Jacob Hall (of the only English ship that was at Vera Cruz) to come to Carolina, where he is free, as all such are; and therefore they call it Puerto Franco. The colonists are now full of pirates' money, and from Boston I hear that the privateers have brought in £80,000.[1]

Nonetheless Lynch was still interested in recruiting de Graff. He noted with some satisfaction that "the *Ruby* met [de Graff] off Cartagena, and I was pleased to hear that the Spaniards noticed how respectful they were."[2] Lynch was happy to have the Spanish believe that the powerful de Graff was in his corner. Soon the governor and the pirate resumed their correspondence.

In late April 1684, de Graff wrote to Lynch to report the incident of the English prize ship he had released:

I present my humble respects and hope that your health is good. I have a few details to give about a small English ship, laden with sugar, which I found in the hands of a Spaniard. I took both ships in the night, kept the Spaniard and set the Englishman free. The English captain told me that the Spaniard was taking him and his ship into Havana, but I gave him the ship back without doing him any harm. I send this short note only to show you that I am far from injuring your nation, but, on the contrary, am anxious always to do it service.[3]

De Graff, undoubtedly chafing at Governor Sieur de Franquesnay's hard line on piracy, was flirting with the idea of coming over to the English side. Sir Thomas Lynch found that idea appealing. In August he wrote back to de Graff:

I have received your letter, and thank you most particularly for letting the poor Irishman go. I shall show my gratitude to you when I have the opportunity, for any one who treats the English well lays me under obligation, and I expect no less from you who hold a patent from the most Christian King [Louis XIV]. . . .

While you behave with such respect to the justice and friendship that exist between the French and the English crowns, I am always your friend.[4]

Lynch and his superiors had already put together a package crafted to bring Laurens to their side. The agreement they had drawn up stated:

Sir Thomas promises a pardon for all offenses and naturalization as an Englishman; but Laurens must take the oath of allegiance and buy a plantation in Jamaica [thus giving him genuine financial incentive to remain loyal to England]. Sir Thomas will also procure the necessary papers for the safe conduct of his wife from the Canaries, provided Laurens pays the fees and the expense of the passage, and he will also procure him the King of Spain's pardon.[5]

All in all, it was a good offer, but before Laurens could act on it, several events occurred that kept him firmly in the French camp.

In April 1684, a month before the offer to Laurens was drawn up, Pierre-Paul Tarin de Cussy arrived in Petit Goâve to replace the unpopular Governor Sieur de Franquesnay. De Cussy found the buccaneers near revolt, many of them deserting Petit Goâve for such congenial locations as the Carolinas, New England, and the Pacific coast of Panama.

The government in Versailles understood that the safety of French possessions in the West Indies depended on the loyalty of the well-armed buccaneers. That was plain reality, and in large part a result of the disaster at Las Aves.

Just as Lynch had recognized the strategic importance of the buccaneers to Jamaica, so de Cussy realized that he could not afford to alienate this quasi militia. He began immediately to retreat from de Franquesnay's strict enforcement of the law. Soon it was business as usual.

About the same time de Cussy was remaking Petit Goâve into a welcome haven for the pirates, de Graff captured a Spanish vessel with dispatches announcing the resumption of hostilities between the French and Spanish, a war that he himself had helped to foment with his attack on Vera Cruz. De Graff left his consorts Yankey Willems and Andrieszoon to blockade Cuba while he returned to Petit Goâve to plan his next move.

De Cussy was as eager to have de Graff on the French side as Lynch was to have him on the English. He greeted de Graff with the respect due a military hero and gave him an honorary commission, a *brevet de grâce*. It was a fine honor for a man known to the world as a notorious pirate. For a pirate who was also a black man, it was astounding. Best of all, it gave de Graff leave to continue to do exactly what he had been doing for the past decade.

De Graff spent most of the summer and fall of 1684 at Petit Goâve. Meanwhile, his subordinates Willems and Andrieszoon continued blockading Cuba.

Off Havana, they intercepted two Dutch West Indiamen, the *Stad Rotterdam* and the *Elisabeth*. Willems and Andrieszoon boarded the ships and discovered that they carried large quantities of Spanish money and several Spanish passengers, including a bishop. The Spaniards had hoped to benefit from Dutch neutrality by shipping their specie and people in vessels that were nominally off limits to privateers. It did not work. The Dutch buccaneers took half of the 200,000 pesos aboard and all of the Spanish citizens.

From Cuba, the two Dutchmen sailed north to the English colonies in America. There they met with the kind of warm reception that was driving Thomas Lynch to distraction. The governor of New Hampshire, Edward Cranfield, wrote to London that "a French privateer of 35 guns [Andrieszoon's ship *Mutine*] has arrived at Boston. I am credibly informed that they share £700 a man. The Bostoners no sooner heard of her off the coast than they dispatched a messenger and a pilot to convey her into port . . ."[6]

Both Andrieszoon and Yankey Willems found a welcome in Boston, especially from a local merchant, Samuel Shrimpton, who was also the wealthiest man in Boston.

The vigorous Puritan ethic in Massachusetts in general, and Boston in particular, had relaxed considerably by this time—at least insofar as business practices were concerned. Boston merchants of these decades are now remembered as pious, frugal, and industrious pillars of the community. While this image may have been true between 1630 and 1670 (and then again after the Great Revival of the 1740s), it certainly wasn't the case during the late 1600s. This becomes evident when we look at how Bostonians were seen by outsiders—as opposed to how the Boston "Saints" saw themselves. Sailors commented about the "sly, crafty tricking designing sort of People" they met in Boston. No one doubted that the city's big merchants were the slyest and craftiest of them all.

One 1699 observer found that "whosoever believes a New-England Saint shall be sure to be cheated; and he that knows how to deal with their Traders, may deal with the Devil himself and fear no Craft." Another noted, "It is not by half such a flagrant sin to cheat and cozen one's neighbour as it is to ride about for pleasure on the Sabbath Day or to neglect going to church and singing of psalms."

Prior to at least the time of the Great Revival, sailors considered Boston a good town for frolicking—primarily because of its "well-rigged" young women—although at least one early-eighteenth-century sea dog warned the amorous young sailor to be careful lest one of these women "give you a doase of Something to remember them by." There were music houses with dancing and entertainment, and good times were to be found at such taverns as the Dog & Pot by Bartletts Wharf, the Widow Day's Crown Tavern by Clarke's Wharf, or the Sign of the Bull by modern-day South Station.

Boston was never as wild as the hell towns of Tortuga or Port Royal. A sailor in search of "a bit of fun" had to be wary and discreet. The Bostonian was more than willing to part him from his cash, but otherwise cared little for outsiders—least of all for those without local property or local family ties.[7]

Andrieszoon set about refitting his ship. It was Yankey's intention to refit once his compatriot was done, but in the interim the king's proclamation prohibiting the aiding and abetting of pirates reached the city. Massachusetts customs agent William Dyre seized *Mutine*. This was not a popular move, and under pressure from Shrimpton and Governor Simon Bradstreet he released her after a brief time.[8] Nonetheless, the buccaneers felt their warm welcome turn cool and sailed for the more temperate Caribbean.

A PIRATE'S WEDDING

Laurens de Graff had been busy as well. Legend has it that while ashore, de Graff met the great love of his life, a Breton woman named Marie-Anne Dieu-le-Veut ("God Wills It"). The story has all the romance of the tale of de Grammont's slaying his sister's suitor, and might be taken with an equal grain of salt.

Marie-Anne was the widow of another adventurer, and was hardly a blushing maiden. In fact, she was at least as bold and headstrong as

Laurens himself. According to the legend, Marie-Anne heard that de Graff had made some remarks about her that she deemed inappropriate. With pistol in hand, she went straight to the tavern to find Laurens and demand public apology. There "the filibuster, filled with admiration for so bold a gesture, proposed he should marry her by way of apology."[9]

With his other wife out of reach behind Spanish guns in the Canaries, and with the moral outlook of a pirate, Laurens apparently had few reservations about adding bigamy to his misdeeds.

Petronila de Guzmán, his first wife, may have been important enough for the English to include in their offer to de Graff, and for the Spanish to use as bait in *their* offer. But as long as Petronila was in Spanish territory, she was as good as dead to him. Pragmatically, remarrying must have seemed the reasonable thing to do, given the outbreak of war.

Marie-Anne was at least as pragmatic. If Thomas Lynch knew about de Graff's first wife, it is a sure bet that Marie-Anne did, too. In any event, the couple were married soon after. The story has it that Marie-Anne wore a brace of pistols with her wedding dress, just the sort of accessorizing one would expect of a pirate's bride.

GATHERING FOR ANOTHER RAID

On November 22, 1684, de Graff finally took his leave of Marie-Anne and Petit Goâve, going to sea with a crew of 120 men in the fourteen-gun Spanish dispatch vessel he had captured the previous spring. After a long passage battling contrary winds, de Graff finally arrived off the Spanish Main. On the evening of January 17, 1685, he came across a squadron of two ships and four smaller vessels.

Believing he had blundered into a Spanish *armadilla,* de Graff cleared for action. An eyewitness aboard de Graff's ship, Ravenau de Lussan, described the near disaster that took place:

> One of those boats on the Eighteenth by break of day, being a Tartane commanded by Captain John Rose, as not knowing us presently, came up and haled us; and as our Captain had a commission from the Lord High Admiral of France, the Count of Thoulouse, we made answer from Paris, and put up our Flag; But

A fourteen-gun fluyt typical of buccaneering vessels

Rose who would not know us so, believing we had no other Intention in feigning our selves to be a King's Ship, than to get clear of him, gave us Two Guns to make us strike, insomuch that taking him really for a Spaniard, we knocked out the head of Two barrels of Powder, in order to burn ourselves and blow up the Ship, rather than fall into the Hands of those People, who never gave us Quarter, but were wont to make us suffer all imaginable Torments, they beginning usually with the Captain, whom they hang with his Commission about his neck. . . .[10]

De Graff's preparations were akin to the philosophy of saving the last bullet for yourself. De Graff knew that he could not trust any promises the Spanish would make to secure his surrender. Pirates frequently prepared for mass suicide when threatened with capture, knowing that the Spanish would give them a far slower death.

The ships maneuvered to within hailing distance before any further exchange of gunfire. De Graff did not yet realize that he had come across the very squadron he was seeking, the ships *Neptune* and *Mutine,* commanded by Yankey Willems and Michiel Andrieszoon.

The smaller, unknown vessel that had initially approached de Graff was commanded by the pirate Jean Rose, who had recently joined Willems and Andrieszoon and who was not familiar with de Graff's ship. Fortunately, Willems and Andrieszoon recognized the vessel and ran up a private recognition signal. With disaster averted, their meeting called for celebration, or, as de Lussan put it, "obliged us to put in at the Cape, and spend that Day to visit one another."[11]

The pirates next headed for Curaçao. De Graff dispatched one of his consorts to request permission to buy masts to replace those that the pirates had lost in a storm. The governor was well aware of Willems and Andrieszoon's plundering of the Dutch West Indiamen the year before and was not interested in aiding the pirates.

The small buccaneer fleet continued along the Spanish Main, meeting with little success in the ventures they undertook. This was the lull before yet another storm. De Graff was, at this point, near the height of his fame and power. He had participated in nearly every major buccaneer raid of the decade and had been the prime mover of several of them, including the sack of Vera Cruz. He had a taste for such action, and was looking for more.

Organization and discipline within the buccaneer community was loose at best. The filibusters tended to come together when it suited them and to go their separate ways when they chose. It was just this fluidity that made them so hard to stop. Infighting did not threaten them, since any disagreement would just lead to a disbanding of that particular group, with an inevitable re-forming of another filibuster army at a later time.

So it was with de Graff's little flotilla. De Graff favored large assaults, but not all in his fleet agreed. They decided to split up. The crews reorganized themselves and the various ships went their separate ways.

In April 1685, de Graff once again presided over a great assembly of buccaneers, this time on Isla de Pinos, or the Isle of Pines, off the coast of Cuba. De Graff's fame was such that when it was known he was planning something, others wanted in. Twenty-two ships gathered at Isla de Pinos, including such pirates as Yankey Willems, the Englishman Joseph Bannister, and Jacob Evertsen, a Dutchman like Willems and de Graff.

The Chevalier de Grammont was also at Pinos. For the past few years, the Chevalier had been somewhat inactive. Following the attack on Vera Cruz, the buccaneers had met at Isla Mujeres, near Cancún, Mexico, and divided their spoils. From there, de Grammont had attempted to sail to Tortuga, but he was plagued by contrary winds. His crew and prisoners were in danger of starvation. Before that happened, he fell on and captured a Spanish vessel, *Nuestra Señora de la Candelaria,* which he robbed of food, five seamen, and most of her sails.

Before setting the Spaniard on her way, he put twenty-two of the prisoners from Vera Cruz aboard her and gratefully issued her master a pass as a safeguard against harassment by other buccaneers, though how effective such a pass would have been is questionable.

At last, despairing of reaching Tortuga, de Grammont sailed for Petit Goâve, where he apparently spent the remainder of the year, no doubt in the company of his old consort de Graff, with whom he had already shared many adventures.

Whether the two buccaneers discussed plans during that hiatus is not known, but when de Graff was once again ready for a big raid, de Grammont was there. As was often the case with such large and democratic—not to mention drunken—gatherings, there was no consensus on the next attack. Many of the pirates favored another move on

Eighteenth-century map of Central America

Vera Cruz, the spoils having been so good the last time. De Graff disagreed. He did not believe that Vera Cruz would be so easily taken a second time.

When it became clear that there would be no agreement among the buccaneers, de Graff left Isla de Pinos and sailed for the coast of Panama.

Rumors swirled about the buccaneers' deliberations at this council. It is possible that de Graff and de Grammont had a falling-out. At least, that was what Sir Thomas Lynch heard from his sources among the Brethren of the Coast. He wrote to the Lord President of the Council, "Grammont is going with his own ship, and four or five more, to Leeward; he has fifteen hundred men and is supposed to have designs on Caracas. Laurens and he are great enemies; but I have heard nothing from Laurens since he took the Spaniard in St. Philip's Bay."[12]

Unable to get their own operation going, de Grammont and the others eventually sailed after de Graff and found him on the Mosquito Coast. It is perhaps a measure of de Graff's status that de Grammont and the others sought de Graff's support, just as they had before Vera Cruz. They seemed to agree that the big operation they were contemplating could not be launched without de Graff.

For once, Lynch's intelligence was not quite on. Whatever the relationship between the two filibuster leaders when de Graff took his leave of Isla de Pinos, it was soon once again a partnership. The idea of a second raid on Vera Cruz was abandoned. Instead, the buccaneers agreed on a new victim.

This time, it was to be Campeche.

31

Of Destruction
and Death

[T]he French are making up a fleet of twenty-two
sail at the Isle of Pines for some design
which is kept very secret.
—Lieutenant Governor Hender Molesworth
to William Blathwayt

APRIL 1685
ISLA DE PINOS, CUBA

PIRATE DEMOCRACY

The decision to stage an attack on Campeche may have started as a secret, but during the summer of 1685 it was common knowledge that the buccaneers were once again gathering for a major strike.

In April of that year, Captain Mitchel of the ship HMS *Ruby*, forty-eight guns, stumbled across the pirates at their rendezvous on Isla de Pinos. Though some of the filibusters who had participated in the sack of Vera Cruz had retired with their loot, many others were back. Gathered on Isla de Pinos were most of the usual suspects—de Grammont, de Graff, Yankey Willems, and Joseph Bannister, among others.

One might expect a certain hostility between the captain of an English man-of-war and a cadre of known pirates, but instead the odd relationship between the pirates and the authorities prevailed. Mitchel does not seem bothered by this large massing of buccaneers, no doubt because he is certain that their target is Spanish. His only concern was that Bannister, an Englishman, should be serving under the French flag. The fact that he was so greatly outnumbered may also have played a part in his accommodating attitude toward the pirates.

Mitchel parleyed with de Grammont. The exchange was reported later by Lieutenant Governor Hender Molesworth, who was serving as acting governor of Jamaica following the death of Sir Thomas Lynch on August 24, 1684, at the age of fifty-two.[1]

He [Mitchel] sent aboard Grammont to know why an English ship was sailing under French colors, and demanded the arrest of Banister for serving under a foreign commission, but they all said

A sixty-gun ship, 1670

that he had not entered the King of France's service, so Captain Mitchel thought it best not to insist further.[2]

In the face of the pirate fleet, Mitchel realized that discretion was the best course of action.

Mitchel may not have been hostile toward the intentions of the pirate fleet, but he certainly was curious, and was willing to speculate. Molesworth wrote:

> Mitchel advises me that the French [buccaneers] are making up a fleet of twenty-two sail at the Isle of Pines for some design which is kept very secret. He supposes it to be against the galleons which were to sail this month from Cartagena for Porto Bello, but I rather suspect it may be against Cartagena itself as soon as the galleons are gone, the place being weakened by detachments from Lima and Panama.[3]

Needless to say, the Spanish were even more curious about the pirates' target. The consensus among the Spanish was that they were planning an attack on Cartagena or Havana. Both ports prepared to stave off pirate attacks. Governor Andrés de Muñibe of Havana reported to the viceroy, "I have raised a parapet of buttresses and mud walls for the defense of the inner channel of this port, which is adjacent to the mouth."[4]

Havana, however, was considered a tough nut to crack, and some of the pirates held Cartagena in superstitious awe. Cartagena was the site of the Monastery of Nuestra Señora de Popa, and many privateers would blame any of their misfortunes on it. In 1669, while he was preparing to raid Panama, Henry Morgan's flagship, *Oxford,* blew up, killing 350 men. Only Morgan and a handful of others survived. The Spanish claimed that on that night, the "Lady of the Monastery" came back with her clothes wet, dirty, and torn. That legend was enough to make many pirates uncomfortable with the thought of sacking Cartagena.

No one seemed to have guessed that the real target was Campeche. This ignorance could have been a great advantage to the buccaneers, had they been able to maintain the element of surprise, but instead they squandered it.

The fleet shifted from the Mosquito Coast to Isla Mujeres. While the main part of the buccaneer fleet lay at anchor at Isla Mujeres, a number of ships were sent to patrol off Cape Catoche, the northeast-ernmost point of the Yucatán Peninsula across a narrow strip of water from the western tip of Cuba. There the pirates kept a watch for other freebooter vessels passing that way that might wish to join in the party.

For a month, the pirates kept up their patrol. It is unclear why they waited so long before staging their attack. They were only three hundred miles from Campeche; it was inevitable that this picket would be discovered. The Spanish coast guard frigate *Nuestra Señora de la Soledad* arrived at Campeche with her small convoy to report that on May 27 they had been chased by an unidentified enemy. Regular reports of pirate activity began reaching Felipe de la Barrera y Villegas, the deputy governor of Campeche. There was even time for Don Felipe to dispatch spy boats to keep an eye on the freebooters and report as soon as they mobilized.

For the buccaneers, it must have seemed like old times. De Graff, de Grammont, Yankey Willems, Michiel Andrieszoon, Jean Foccard, and Pierre Bot had all been together at Vera Cruz, Caracas, and Mara-caibo, had all sailed together before, had all plundered the Spanish, side by side. Many of them had been there on the sandy beach of Las Aves with the Chevalier de Grammont. There were relative newcom-ers as well, such as Rettechard and Jacob Evertsen.

THE SACK OF CAMPECHE

In late June, word arrived that the pirates were under way, rounding the Yucatán and sailing south toward Campeche. After decades of pirate raids, the Spanish were quite adept at making valuables and noncombatants disappear into the countryside, even at a moment's notice. With several months' notice, Campeche was cleaned out.

On the afternoon of July 6, 1685, the pirate fleet arrived. It con-sisted of six large ships, four smaller ones, six sloops, and seventeen piraguas. From these vessels came a force of seven hundred buccaneers manning a flotilla of ships' boats pulling en masse for the city's landing. The people of Campeche were far from surprised.

Ready for action, de la Barrera sent out four militia companies totaling four hundred men to cover the beach where the pirates intended to land. Though the militia were outnumbered, they still held the advantage. In any amphibious operation, be it a pirate landing in Mexico, or the Allied landing on the beaches of France, seaborne troops are at their most vulnerable as they are attempting to disembark. The freebooters understood that. In the face of the militia, they drew off and did not attempt to land.

The buccaneers spent the night in the open boats, bobbing out of range of the Spanish defenders' guns. The next morning, they took up their oars and headed back toward their ships, giving every appearance of abandoning the attack. One can imagine the cautious relief felt by those remaining in Campeche.

The relief was short-lived. Before the pirates reached their ships, they sheered off and pulled hard for the edge of the city, coming ashore before the militia could realign themselves to oppose their landing. Rather than hurling themselves immediately upon the city, the pirates formed up into four orderly columns and advanced on different points of the city's perimeter. One hundred men formed the vanguard under Captain Rettechard, and two hundred under de Graff headed straight for the center of Campeche, while another two hundred led by Jean Foccard marched down a street parallel to de Graff. De Grammont led two hundred more in a march that circled around the city.

It was brilliant organization, brilliantly executed. The Spaniards fell back in the face of this onslaught, overwhelmed by the tactics and the sheer numbers of the invaders.

On Governor de la Barrera's insistence the coast guard frigate *Nuestra Señora de la Soledad,* which had first brought news of the pirates to Campeche, had remained in the harbor to help with the city's protection. Her captain, Cristóbal Martínez de Acevedo, intended to scuttle the ship if it appeared she might be taken by the buccaneers.

Seeing how quickly the city's defenses were falling apart, Martínez decided that scuttling was too slow. He ordered a trail of gunpowder run to the ship's magazine. After the crew abandoned ship, Martínez lit the fuse from the ship's boat.

The *Soledad* blew up with a tremendous explosion. The blast not only destroyed her but utterly unnerved Campeche's defenders and shattered their morale. They abandoned the fight and took refuge in various strongholds around the city, primarily the citadel. Governor de la Barrera's attempts to rally the defenders were to no avail. He and the men under him were driven out of the city and into the country-

Buccaneers attacking a town

side. The governor abandoned his attempt to save the city and saw instead to the safety of his wife and children. Campeche was in the hands of the buccaneers.

The pirates spent the next few days routing out the troops who remained in the various strongholds. Finally, only the citadel remained. At dawn on July 12, 1685, the pirates began bombarding that fortress. Their attack was interrupted a few hours later by the appearance of two relief columns of Spanish militia, which had marched down from Mérida de Yucatán, the provincial capital about ninety miles away.

The Spanish troops, apparently, were quite confident of their ability to drive the pirates back into the sea. When de la Barrera tried to assemble the troops in a nearby town to mount a coordinated attack,

De Grammont at Campeche

he was ignored. Instead, the troops marched right on in a headlong assault.

The elaborate defenses that de la Barrera had built up to defend Campeche against the pirates now helped the pirates defend Campeche against the relief force. The Spanish troops marched right into well-aimed volleys from the pirates, who shielded themselves behind Campeche's ramparts. The Spanish were mowed down as they approached the walls of the city. The casualties were too many to count.

The Spaniards fought harder, but not smarter. They flung themselves again and again at the city's defenses, into the pirates' devastating fire. The battle raged for the better part of the day, until finally de Grammont led his troops in another flanking maneuver that caught the Spaniards in a crossfire. That was enough for the militia. They withdrew from the field, abandoning the city to the freebooters.

The Spanish garrison in the citadel had been holding out waiting for

An attack on a Spanish citadel

the relief column to arrive and drive the buccaneers off. The pirates' victory quashed any hope they might have had. Threatening to shoot any officers that interfered with them, the troops in the citadel deserted the place, slipping away in the darkness.

By eleven o'clock, the citadel was abondoned. A couple of English prisoners, freed by the fleeing garrison, called out to the pirates that the way was clear. The pirates feared a Spanish trick and ordered the Englishmen to discharge all of the cannons so the pirates could be certain they were advancing in the face of unloaded artillery. When this was done, the old freebooting partners, the Chevalier de Grammont and Laurens de Graff, personally led the men over the wall and into the citadel. Campeche was entirely in their hands.

The attack on Campeche was one of the few instances in which filibusters encountered a fully prepared enemy and attacked anyway. It says much about the courage of de Graff and de Grammont and much about their skill. No wonder colonial governors were eager to bring such men under their direction.

Once the city was in their hands, the pirates settled in for a long stay, just as they had in Maracaibo seven years before. Once again, de Grammont organized an ad hoc cavalry and dispatched mounted buccaneers to the countryside to plunder as they might. Pickings were slim. With all their forewarning, the inhabitants had cleaned out Campeche of anything the buccaneers might want. The pirates' frustration mounted.

Two weeks after the taking of Campeche, de la Barrera received word from the governor of Yucatán, Juan Bruno Téllez de Guzmán, that all available troops were to be assembled at the town of Tenabó, about twenty-five miles from Campeche. De la Barrera accordingly set out with the handful of men he had left under his command, only to be captured by a patrol of de Grammont's mounted buccaneers. De la Barrera once again entered the city of Campeche, not in triumph but as a prisoner.

FRUSTRATION AND FALLING OUT

During the nearly two months the buccaneers held Campeche, they carried out the customary atrocities of the time, which, in this instance, were aggravated by their frustration at finding so little plunder.

August 25, 1685, was the feast day of the patron saint of Louis XIV, and the French filibusters celebrated in high style. The next day they made preparations to get under way, possibly prompted by an outbreak of fever. Before the buccaneers abandoned a city, they would usually exact a ransom in exchange for not burning the place to the ground. In the case of Campeche, it looked as if such a ransom would be the primary source of any loot they might garner.

A message was sent to Governor de Guzmán demanding eighty thousand pesos and four hundred head of cattle for the protection of Campeche. In return, the governor sent a bluntly callous letter, assuring the pirates that "they would be given nothing and might burn down the town, as [Spain] had ample funds with which to build or even buy another, and people enough with which to repopulate it."[5]

Such insolence would have angered the pirates in any event, but coming at the end of two fruitless months, it pushed them over the edge. De Grammont ordered a number of houses put to the torch by way of demonstrating his sincerity. The next day he sent another message inland, this time promising to begin executions if his demands were not satisfied. The governor sent the same reply as before.

De Grammont did not make idle threats. The day after receiving the second message from de Guzmán, the Chevalier paraded his prisoners in the town square and began to systematically execute them.

After half a dozen men had been hung, Felipe de la Barrera y Villegas and a few other leading citizens, who lacked nothing in courage and honor, approached Laurens de Graff, whom they considered the more humane of the two buccaneer leaders, and tried to bargain for the lives of the prisoners. De la Barrera and the others had nothing with which to bargain, save their own lives. They offered themselves as de Graff's slaves if he would spare the prisoners de Grammont was determined to execute.

De Graff was not interested in having the Spaniards as slaves, but neither was he interested in continued bloodshed. He and de Grammont had a lengthy tête-à-tête. There is no record of what was discussed or what motivated de Graff to take up de la Barrera's cause. Perhaps he had no stomach for senseless slaughter. Perhaps he remembered how he had mortally wounded Van Hoorn after the sack of Vera Cruz, when it was Van Hoorn who wanted to put helpless prisoners to the sword.

The end result was that the lives of the prisoners were spared. The buccaneers abandoned Campeche after spiking the guns and carrying off many of the prisoners for ransom. It was early September 1685. The pirates had spent two months at Campeche and had little to show for it.

After rounding the Yucatán Peninsula once more, the buccaneers stopped at Isla Mujeres, where they divided what scant loot they had. Then the filibuster army split up.

It appears that the understanding reached between de Graff and de Grammont concerning the prisoners at Campeche was not entirely amiable. There may have been friction between the two men going back to the gathering at Isla de Pinos, where de Graff had sailed off in frustration. Whatever the cause, these foremost of the buccaneer leaders went their separate ways.

As fate would have it, they would never see each other again.

DEATH OF A *FLIBUSTIER*

Pierre-Paul Tarin de Cussy, de Pouançay's successor, worked to bring the buccaneers into the official French government and military. In September 1686 he commissioned the Chevalier de Grammont "lieutenant du roi" of the coast of Saint-Domingue. At the same time, de Graff was made a major, though his title would later become "Laurens-Cornille Baldran, sieur de Graff, lieutenant du Roi en l'isle de Saint Domingue, capitaine de frégate legère, chevalier de Saint-Louis."[6]

De Grammont, no stranger to high office, was not quite ready to settle down. He had in mind at least one more attack on the Spanish. Where his former lieutenant Thomas Paine had failed, de Grammont was ready to try his luck. His target was St. Augustine.

Eight months after quitting Campeche, de Grammont was sailing in company with Nicholas Brigaut and another pirate from the Campeche raid. In April 1686 they stood into the Atlantic and came to anchor south of St. Augustine under Spanish colors. Brigaut, sailing a small Spanish vessel they had captured, went on ahead to scout out the attack. De Grammont and the other vessel waited, but Brigaut did not return.

At last, de Grammont decided to go himself, no doubt hoping to discover what had become of his consort. As it turned out, Brigaut's ship had been lost in a storm, and de Grammont sailed right into the same foul weather. The gales drove him north along the coast. He was never heard from again.

It was not for another year and a half that any word of de Grammont reached French ears. After escaping from a Spanish prison, a buccaneer named Du Marc passed along a rumor to de Cussy that de Grammont's ship had been lost with all hands.

After all the danger de Grammont had faced, after multiple shipwrecks, after a near-fatal sword wound, after the sacking of Maracaibo, La Guaira, Vera Cruz, and Campeche, the sea ultimately claimed him.

The life of the Chevalier de Grammont was perhaps closer to the Hollywood vision of a buccaneer than that of any other pirate in history. Fleeing France under a cloud, he rose to lead armies of the most dangerous and undisciplined men, and commanded either the fear or the respect of officials of every nation represented in the West Indies, only to die an anonymous death. One of the great buccaneer captains of all time was gone, and one of the wildest chapters in the history of the filibusters had come to an end.

32

Of Men-of-War and Pirates

OCTOBER 29, 1998
LAS AVES

Life aboard the *Antares* was good, for the most part. We were pleased with the work we were accomplishing. Mike Rossiter was happy with the footage he was shooting. Every evening after dinner, Eric Scharmer would review the underwater footage on the boat's VCR and we would gather around and evaluate our discovery work.

The vessel itself was a little cramped. As the days went by, we started to get a bit tired of one another, and it started to seem even more cramped. That is inevitable on an expedition such as this.

There were other problems, too, both with accommodations and group dynamics. Margot and I had a private cabin into which the holding tanks for the heads seemed to vent. The smell made me sick. Chris and Cathrine also had a private cabin, which didn't smell much better.

Todd, Carl, and Eric were bunking together. Todd and Carl yucked it up in the evenings like a couple of high school football players on their first away game. It was the Special Forces/SEAL rivalry.

Rivalry, of course, is a part of the game. Todd and I have been at each other for more than twenty years. From the minute we meet in the airport to the minute we get home, it is an endless competition of teasing and ribbing to see who is going to crack first, who is going to lose his cool. Believe me, on an expedition you must have a thick skin to endure some of the comments thrown at you. Everything is fair game, nothing is sacred: from pointing out someone's lack of physical prowess to questioning the fidelity of his wife or girlfriend. The only object is to get a reaction out of the other guy.

This is always a lot of fun, at least for the first few days of the trip. After a while I am ready to knock it off. But Todd will never quit. He never lets up. By the end of each expedition I am ready to crack. But after a month or two away from him, I'm ready to start the psychological jousting all over again.

This time I had Carl as a secret weapon. I could get Carl and Todd going on each other, and that would lighten the load on me. This worked for a while, but finally, as we neared the end of the trip, I had to tell the two of them to lighten up. They were driving me and everyone else nuts. Todd just smiled at me and my tacit admission that, for once, he had won.

Charles was, predictably, a problem. I believe that he considered himself the leader of the expedition, and that he felt he had to demonstrate his authority. Sometimes he would come with us while we were mapping wrecks, and sometimes he would be off with his own people doing God knows what. It was like having a shadow expedition.

Charles would not let up on his idea of picking a wreck site and going over it with the same eye to detail he had once used as a dentist. It was annoying because my team was doing straightforward underwater surveying, working like a well-oiled machine, mapping wreck after wreck, just as we had hoped. I wanted an underwater portrait of the disaster as a whole, and I did not want to slow down the pace.

Besides, I knew what Charles really wanted to find. I had been told that his modus operandi was to use science as a cover for his goldmining ventures in the Amazon jungle. I had seen enough to realize that Charles was like a reincarnated conquistador. As one journalist later put it: "Instead of a cross, he and his scientist-missionaries carried cameras, computers, and blood-sampling equipment into the wilderness as a polite prologue to the engineers of destruction who always followed on their heels."[1] In the case of Las Aves, the ex-dentist turned naturalist turned anthropologist was now shedding his skin yet again to turn underwater archaeologist. In all his incarnations, however, I believe there was one common denominator—gold.

Charles and I had more than one argument about the project mission, but we weren't in the jungle; we were in my element now, and "gold," as my uncle Bill once said, "don't float."

Mike Rossiter and Charles got into it as well. Their relationship that had started badly only became worse. Mike would tell us that we needed to tape one thing, and Charles would insist that we do something else. It was the same thing that had happened in the planning stage, with Charles trying to dictate the mission of the expedition. I was ready to move north along the reef and locate the next wreck, and Charles was talking about renting a plane for aerial reconnaissance or some dammed thing. Much later, I would learn that he was not a stranger to faked flight plans.[2] I'm sure it bugged Charles that, in the end, Mike called the shots because he was paying for it all. And Mike looked to me to lead the expedition.

Charles had his own boat, but he was always aboard the *Antares.* That wasn't a problem. For the most part we kept our opinions and feelings to ourselves and maintained a cordial attitude. It wasn't even a problem when Charles expressed an interest in moving aboard. But soon he was insisting that his crew move aboard, too. The *Antares* could not hold that many people. Todd and Ron had carefully figured how much food and water would be needed for the expedition based on the number of team members I was bringing. With Charles and his entourage, we would run out of supplies long before we were done. There was no way that Charles could move aboard the *Antares,* and he was told as much. That did nothing to ease the mounting tension.

Adding to the mix was the other third of our expedition, Max Kennedy and his friends whose base was their own vessel. One of the worst mistakes made by some academic archaeologists is to either completely dismiss the aid of what they call "avocationals" or exploit them as slave labor. Men like Max and his friends bring a freshness and enthusiasm to projects that remind me of why exploration is so appealing in the first place. They had to get back to their families and their jobs after the first week, and I was sorry to see them go.

The food aboard *Antares* was great at first. There were eight crew members on board the boat, including a cook. The cook was used to catering to American and European sports divers, so the meals he made were slightly ethnic but not too much so.

Soon, however, the supplies started to run out and the meals became a greasy mess. That concerned me. Food is really important to my crew, and they are always hungry. Taking someone's dessert is

Barry Clifford and Max Kennedy

considered nearly as serious as stealing someone's girlfriend. I've seen fights nearly break out over the last piece of chicken. After a hard day of diving, a good meal is essential, and we were not getting them anymore.

Perhaps worst of all, the air-conditioning no longer worked. In the heat we experienced down there, functional air-conditioning was important. Without it, there was no escaping the heat, not even in the water. As bad as the surf was over the reefs, on the day that we went to look for the *flibustiers,* it was the heat that came the closest to killing me.

33

Flibustier

The heat was brutal. At the surface of the water the temperature gauge on my dive console read in the high eighties. That's *in* the water. Twenty feet below the surface it still felt like you were taking a bath. When we had to work on deck, we would constantly jump off the boat just to cool down, but it hardly helped at all.

Most people would not think it, but a diver can overheat, working hard and fast and having to wear a full wet suit because of the coral. With the Aga it's easy to skip-breathe, meaning that you will actually hold your breath for a beat, and then resume breathing. When that happens you get a slight carbon dioxide buildup, which can cause dizziness and even blackouts.

The sharks that Ron had warned us about never did make an appearance, except for a nurse shark we saw sleeping on a ledge. But the reef was teeming with barracuda, huge barracuda, five or six feet long, the biggest any of us had ever seen. As we worked, the barracuda would hang back in the shadows and watch us.

Barracuda have mouths that bristle with razor-sharp teeth. They always seemed to be watching. They can hang in the water the way a dragonfly treads air. One of the crew on Max's boat told us that on a previous trip to Las Aves, they had caught a barracuda on hook and line. While trying to land it, it bit the bottom step off their dive ladder.

We learned a lot about the barracuda from the indigenous people who know the area, in particular the conch divers. Those men were in the water at the same time we were, and we took some comfort from that. They told us that a barracuda will change colors and circle around its prey before attacking. We kept an eye out for that behavior, but we didn't have any problems.

The conch divers are always on the lookout for sharks, but every so often a diver disappears. Sometimes the conch divers will spear barracuda, and on occasion our cook would buy one from them and serve it up for dinner. They taste great, though they can be dangerous, even when they are dead. Barracuda eat fish that eat coral, and the coral that the smaller fish eat can make a person terribly sick. We took our chances and never had a problem.

Actually, the locals had more problems than we did.

One day a couple of fishermen came by the boat. One of them was cradling one hand in the other and bleeding all over the place. He had been holding a barracuda by the gills, which are sharp, and his

A dangerous dinner

hold slipped and the gills gashed his hand open. The skin was sliced right open and a big patch was hanging in a **V** shape. They asked if we could help.

Since Todd Murphy is an army medic, he took a look at the wound. It was bad, more than Todd felt he could handle. The fisherman needed to get to a doctor. Todd asked the diver who had brought him over what they would do if the hand could not be sutured onboard. The man said they would put a bandage on it and send him back to work.

Given the choice between his field suturing or a bandage, Todd knew the fisherman was better off with field suturing. He put twenty-five stitches in the man's hand. In the end, both were pleased with the results.

The barracuda were not our only visitors. We were boarded nearly

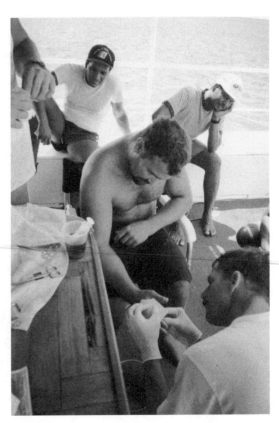

Todd Murphy examining a barracuda cut on a
fisherman's hand

every day by either the navy or the coast guard. It was very nerve-racking, not knowing what they would spring on us next. We were never certain whether or not they knew that we were filming. It was anxiety-provoking. The permit question was like the Sword of Damocles. We were waiting for the morning when we would be shut down for real, or worse.

Mike Rossiter kept up his barrage of radiotelephone calls. The contact with the Venezuelan ambassador was paying off. The ambassador had mobilized his assistant in Caracas, and she was trying to find out what was happening and what we would need to get our permits recognized.

Halfway through the expedition, the worst happened: the wind came back, that damned wind.

The morning of the fifth day on the site it came from the east, setting up a line of breakers along the reef. Fortunately, it was not nearly as bad as it had been the first time around. Wind has to blow steady and strong for at least a few days before big seas build up. If the wind did not let up, it could become bad.

We had worked our way north along the reef, exploring and mapping the wrecks of the *Hercules,* the *Défenseur,* the *Prince*. It was an awesome sight, and more so when you looked at the wrecks and considered the terrible tragedy that had put them there.

At one site we found two anchors in an odd configuration. The stock (the wooden crosspiece) of one apparently had been removed, and through the anchor ring another anchor had been inserted. This created a double anchor, almost like a grappling hook. The men aboard that vessel must have rigged the anchors up that way in one last desperate attempt to get the flukes to hold the bottom and to keep the ship off the reefs. There is no historical source evidence that the ships had enough warning before striking to try to get anchors down, but perhaps they did. I couldn't think of another reason for the two anchors to be joined in that manner. There, encrusted and motionless on the bottom, was vivid evidence of the French sailors' last desperate attempt to save their ship. It was a futile effort, for there was no place for an anchor to bite on that hard coral bottom.

Very often, we would find anchors farther out to sea. If we followed a straight line from the anchor to the reef, there we would find a wreck.

Just south of the big freighter that had driven up on the reefs was another wreck, a small ship, a coastal merchant vessel. As it ran over

Barry Clifford with two anchors on the reefs of Las Aves

the reef in its final death throes, it smashed down a swath of staghorn coral, damaging the reef but clearing a narrow path for us to the open water. A rusting thirty-foot section of hull was all that was left of the ship. Approximately three hundred yards south of the wreckage was one of the wrecks marked *flibustier.*

Before the seas became too high, Margot and I went out to the site and had a look around. We went by ourselves. We wanted to be alone for a while.

We found a wreck, right where the map indicated, in just six feet of water. There were the telltale ballast stones, and a scattering of cannons. Not huge cannons, like those you would find at the site of one of the big men-of-war, but smaller, appropriate to the size of a buccaneering ship. And, most intriguing of all, we spotted an encrusted shape that looked very much like a crate or chest of some kind.

That was all there was. It was not much to look at. Still, it occurred to me that I might be, for the second time in my life, looking at the remains of a pirate vessel. If that was true, then it would make us the first modern explorers ever to set eyes on two pirate wrecks. We swam, knowing that for a moment we were the only people to know

Diver inspecting a cannon

its secret location. During the long Cape Cod winters to come, this moment would be recalled again and again.

The next day Margot and I went back, hoping to get a better look at the wreck. D'Estrées' map showed the *flibustier* as the wreck closest to the island. I envisioned d'Estrées standing on the beach looking out over the reefs and the shipwrecks sitting on top of them. Since the pirate ships would have been the closest to him, it seemed reasonable to me that of all the shipwrecks, the pirates' would have been the most accurately located on the map. By finding them and comparing them to the map and the overlays we had done we could establish an accuracy factor for the entire map.

I had the other team members scour the reef from the wreck Margot and I found to the edge of the island. I had to be sure there was no sign of another wreck between the presumed pirate wreck and the point of land where the island began. If there were no other wrecks between, then we could assume that the wreck we had found was the *flibustier*, given the available evidence.

The winds had been building all night, and now the waves were getting high and the going was treacherous. We took the small chase boat out to the wrecked coaster and anchored it in the coaster's lee, using the hull of the old ship as a windbreak. It was over-the-reef time, just as before, and we weren't looking forward to it.

We would snorkel, as the scuba gear was in the chase boat outside the reef. We went over the side of the boat and into the water. We paused behind the wreck of the coaster, orienting ourselves and psyching ourselves up for another fight with the reef. Then we left the shelter of the wreck and plunged into the surf. It was like stepping into a blizzard.

We began to work our way over the reef, pulling ourselves along through that terrible current that I remembered all too well. We grabbed the coral to pull ourselves along, but the coral was brittle and it would break if we applied too much pressure.

The tropical sun was beating down on our dark wet suits. I had taken a black nylon shirt, a chafe shirt, normally worn under a wet suit, and wrapped it around my head to keep the sun off. That was a mistake. The sun on the black shirt only made my head hotter. Instead of kicking my legs and pulling myself along with the coral, I tried to swim, which turned out to be another mistake. The current was very strong, and the effort it took to swim against it was physically draining.

Despite being in the water, we were getting overheated. Our wet suits were wicking the moisture away from our bodies.

We fought our way out over the reef, and by the time we got to the open ocean we could barely see the chase boat because of the high seas, and that meant the people in the boat could not see us. It was several hundred yards away. It had been our plan to swim back after exploring the wreck. Normally that distance would have been no big deal, but now it seemed nearly impossible. What I did not realize was that I was suffering from sunstroke.

We had exhausted ourselves just getting to the wreck site. We were not able to accomplish much. We decided to head back in. But how?

We were not sure if we had the energy to swim to the chase boat outside the reef. We were wearing weight belts, and it would have been possible to drop our belts and surf back over the reef, floating in our wet suits. But the path through the coral was down current, and we would have been really cut up or worse. And I was not ready to drop my weight belt. Doing that is an admission that you are in big trouble, and I was not yet willing to admit that to myself.

Going back over the reef was not a good idea. I could see Carl and Todd in another boat on the outside of the reef about a half mile away, so we started swimming for them. That's usually not far for us, but there was a bit of a current holding us back. It was becoming difficult to breathe. I looked at Margot and could see the strain on her face. I was surprised at her composure; it was the closest I'd ever come to losing it all. The sun was baking us in our wet suits, and our strength was going fast. I thought that I was strong enough to make it, but I wasn't sure about Margot, and I knew that I did not have the juice left to pull us both to the chase boat.

She later told me that she was planning to pull me in if she had to. I believe she would have, too.

We waved, but the crew aboard the chase boat didn't see us. We swam toward it, but with each stroke we became more overheated. I was starting to hyperventilate. I unzipped my wet suit and forced myself to calm down. We were both on the edge of blacking out.

And then, at last, the crew of the chase boat realized we were in trouble. They pulled anchor and raced over to where we were. I was playing it cool, as if there was no real problem, but I honestly don't how much longer we could have kept swimming.

Once aboard, Margot became sick from exhaustion and *mal de mer* brought on by the choppy conditions. She was so sick, in fact, that she asked me in all seriousness if I thought she was going to die. Her tone was resigned and a bit apologetic, as if she felt bad for spoiling the party. I said, "No, you won't die, though you may feel so bad that you'll wish you would." She was in rough shape for the rest of that day.

I wanted to collapse in the bottom of the boat, but I was not going to give Todd the satisfaction of seeing me prostrate. Instead, I sat and let my breathing return to normal. Soon I was ready to go in again.

I am not a philosopher, nor am I an expert on interpersonal relationships. But something I have learned to value, above all else, is the importance of mutual commitment—of having someone you can "ride the river with," as the old-timers used to say out in Colorado. This experience made me realize that Margot was the proverbial "girl for me," and so I asked her to marry me at midnight on the last full moon of the millennium. To my astonishment she accepted my proposal.

The rest of the team had finished combing the reef from the wreck site to the point of land. We did visual searches and sweeps with metal detectors, back and forth. After searching every inch of coral and finding no trace of another wreck, it was clear that the one we found was the closest to the beach, and hence the *flibustier.*

Chris and the others were on the wreck now, measuring and plotting the artifacts, and the film crew wanted to get some footage. I went in with dive gear this time, Aga and tanks, which was much less exhausting. I searched for the chest, but I was unable to find it again. Perhaps it was my imagination, but I don't think so.

The next morning we woke to find the wind was back. We prayed it would not increase. Unfortunately, it continued to build, day after day, blowing steady and increasing by four or five knots every day. The seas started getting bigger along the reef, and the currents came back. Chris and I recalled how it was impossible to work outside the reef when the wind blew. As the seas continued to build, Charles hit a wave wrong while going over the reef in his Boston Whaler and the boat flipped end for end. Luckily no one was hurt. But it was a sharp reminder that the weather could shut us down as fast as the navy. With each extra knot of wind, we felt the pressure to finish grow more acute.

By the fifth day I had the flu. Diving with the flu is pure misery. With plugged sinuses it is hard to equalize the pressure in your ears, and so they hurt like hell. I won't describe the unpleasantness of sneezing into the face mask of an Aga. Since our time on the site was so limited, there was no time for the luxury of staying in bed to recover.

Margot and I had explored one filibuster ship the day before, and I wanted to get a look at the second one. I juxtaposed d'Estrées' map with a 1940s aerial photo that Charles had to get a general idea of where the wreck should be before heading out to the site. D'Estrées' map was proving to be accurate, but from the beach d'Estrées could not get as clear a sense for the shape of the island as could someone flying overhead. In other words, his indication of where the wrecks were was correct, but his drawing of the island was off.

No doubt, things had shifted in three hundred years. That was one of the problems I had had with *Whydah,* and one of the lessons I had learned. Cape Cod today is not exactly where Cape Cod was three centuries before, nor is Las Aves. It took a little calculating to translate the position indicated by the admiral to the corresponding position on the reef.

This was a special dive for me, another pirate ship. I waited until the others had gone off to a wreck site in the north. Charles followed them—as I knew he would. Margot, Ron, and I then headed off in a small skiff by ourselves.

We motored across the lagoon. The wind had temporarily dropped off, the seas were placid, and we were able to put the boat right at the location I thought d'Estrées had indicated. We stood in the bow of the boat and looked down, down through the pristine water to the bottom, a canopy of mottled blues and browns and whites.

There, just below us, encrusted with coral but still unmistakable, was an anchor. Once again, the French admiral was spot on. It was that simple.

Margot and I went over with just snorkels. The water was shallow and clear and ideal for snorkeling. We floated on the surface and looked down at the bottom, then kicked our way to the bottom to get a closer look. I had seen something that I hardly dared hope was true. Just as they had been stowed down in a pirate ship's hold more than three hundred years before were three barrels right in a row. A few feet away lay two more.

Back on the surface I told Margot I had seen what I thought were barrels. She had seen them too, and she could hardly contain herself. I enjoyed seeing her so excited. Barrels do not generally last three centuries. Of the hundreds and hundreds of artifacts we observed at Las Aves, those were the only barrels. The barrel was the universal means of storing things. There could be almost anything inside them.

We swam out to the southernmost point of the island of Las Aves, searching along the reef for more wreckage, but we found none. Later, we returned to the wreck of the filibuster ship and those barrels.

When I say that we saw barrels, that statement needs qualifying. Like most things that have been underwater for so long, they no longer looked very much like barrels. They looked like three uniform lumps of coral, one right next to the other, and more or less the size and shape of barrels. Given their size and shape, and their position one next to another, corresponding to the way barrels were stowed in the seventeenth century, I felt confident about what I was looking at.

I decided to try a metal detector on them. It was still possible that they were just lumps of coral, coincidentally arranged. If the metal detector indicated no metallic content, they might be barrels filled with crockery or meat or water or dried peas or any of the hundred nonmetallic things that eighteenth-century sailors stored in barrels. It

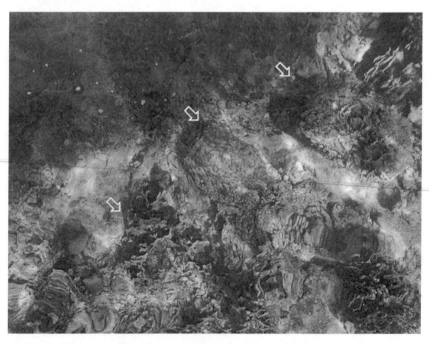

Three intact barrels on a pirate wreck

was also possible that they were not barrels at all. There was no way to tell. If, however, the metal detector showed the objects were "hot," then they had to be barrels.

It was with some trepidation that I went down with the metal detector. I was excited about the discovery and I did not want to be let down.

I swept the White detector over the barrels. They were hot. The needle jumped and the buzzer buzzed in my ear. Metal. They were barrels for certain.

What was in them I do not know. It could have been gold or silver. It could have been musket balls or nails. The only way to know for sure was to raise them and open them. As much as I wanted to do that, I refused to abandon my principles—though my alter ego was screaming in my ear to open them!

What I did know was that this was the finest moment of the expedition for me—regardless of the contents of the barrels. Based on d'Estrées' map, and the artifacts we had seen, our team had found what appeared to be two more pirate-ship wrecks. While their exact identification may—or may not—be established by future archaeological work at this environmentally sensitive site, I am satisfied that these wrecks represent filibusters from the very beginning of that "golden age of piracy" that would ultimately produce Sam Bellamy and the *Whydah*.

There had been three confirmed pirate-ship wrecks ever found, and I was there.

34

The Battle at Alacrán Reef

*[De Cussy] told me that the French King had
made Grammont (whom we took to be lost) his second
lieutenant, and Laurens his third major.*
—Lieutenant Governor Hender Molesworth
to William Blathwayt

SEPTEMBER 1685
GULF OF MEXICO

After the sack of Campeche, de Graff set sail in his ship *Neptune* in company with four other buccaneer captains, beating their way east against the trade winds. On September 11, the pirates sighted to windward of them a powerful antipiracy squadron of the Armada de Barlovento. The Spaniards immediately gave chase.

The armada, under the command of the aged Admiral Andrés de Ochoa, had been at anchor at Cartagena in early August when word reached it of the sacking of Campeche. The armada was ordered to sail at once, find the pirates, and punish them. One of the captains under Ochoa was Andrés de Pez y Malzárraga, the young captain who had been so humiliated by de Graff at Cartagena. He undoubtedly was thirsting for revenge.

Ochoa searched the Cayman Islands and Roatán before intercepting de Graff off the Yucatán. On seeing the Spaniards, de Graff and his consorts fell off and ran for it, being greatly outmanned and out-gunned.

Pierre Bot's ship *Nuestra Señora de Regla* and another proved to be the slowest of the five pirate vessels. As the chase continued, they fell farther and farther behind. Bot began to jettison whatever he could to lighten *Regla,* starting with three large canoes he had stolen at Campeche.

After four hours, the Spanish vice-flag ship, *Nuestra Señora de la Concepción,* closed to within range of her great guns. Her captain, Antonio de Astina, and Bot exchanged furious broadsides. It was a lopsided fight. *Concepción* was more powerful than Bot's *Regla,* even without the aid of the other powerful Spanish ships coming up to join the fight. There was no escape. The French buccaneer hailed the Spaniards, and offered to strike if they would grant quarter.

It must have been a terrible decision to make. Despite whatever promises Bot might secure from the Spanish in the heat of battle, he could not have been confident that they would live up to them. He knew the hatred he and his kind had inspired.

A boarding party from the *Concepción* took possession of Bot's ship. They began to loot it shamelessly, like pirates themselves, despite the officers' attempts to keep their men under control. Weapons and valuables were pilfered. The situation grew worse with the arrival of a boarding party from the flagship, *Santo Cristo de Burgos,* who tried to outpillage their rivals from the vice-flag ship.

The Spaniards found 130 buccaneers, along with more than thirty captives from Campeche and the booty taken by Bot and his men. When the looting was done, the officers recovered no more than thirty pounds of ornaments stolen from Campeche's churches and a few coins.

The Spanish squadron was not able to close with de Graff that day, but the following afternoon he was spotted again by the Spanish frigate *Nuestra Señora del Honhón* near Alacrán Reef, about eighty miles north of the Yucatán Peninsula in the Gulf of Mexico. The eight-gun vessel *Jesús, María y José,* under the command of Andrés de Pez, was sent to beat back to windward and advise the admiral. It was not until four o'clock the following afternoon that the lumbering flagship and vice-flag were able to run downwind and locate *Neptune.*

De Graff was effectively trapped. The Spaniards had the weather gauge, meaning that they were upwind of his position and could drop down on him, while he was hard pressed to tack up to them, and harder pressed to get past them to windward. It was easiest to continue to flee downwind, but eventually he would pile up on the coast of Mexico. The Armada de Barlovento at last had Spain's "Public Enemy Number One" right where it wanted him.

As the two big Spanish ships closed with him, de Graff tried desperately to work his way to windward of them, jettisoning everything he could, just as Bot had done. It was no use. There was no passing the big Spaniards to run away upwind. Night fell before the battle was joined.

At dawn on September 14, the Spanish opened up on de Graff and *Neptune,* and the buccaneer returned fire. The fight lasted all day long, with *Neptune, Santo Cristo,* and *Concepción* exchanging broadsides, circling and maneuvering for advantage. The Spanish ships were big and powerful, but they were also slow and clumsy, and de Graff handled *Neptune* brilliantly. All day long, he evaded the killing broadsides of the Spanish with their superiority in weight of metal.

The Spanish flagship fired fourteen full broadsides into *Neptune,* and *Concepción* expended sixteen hundred rounds. This was met by devastating fire from *Neptune*'s great guns and muskets. For all the iron the Spaniards hurled at de Graff, they managed only to shoot away a few of *Neptune*'s spars, none of which crippled the pirates' ship.

By dusk, there was still no winner in the exhausting battle, thanks to de Graff's uncanny seamanship and gunnery. Admiral Ochoa was not well. Too weak to stand, he had directed the battle from a canvas chair on his quarterdeck. As the sun began to set Ochoa's condition worsened, and he was given the last rites. Command of the squadron was turned over to Vice Admiral Astina of the *Concepción.*

De Graff was still fighting for his life. In the darkness, he threw everything over the side, including his cannons. This was the last toss of the dice. He gambled that rendering himself defenseless would allow him to claw to windward of the big Spanish ships. If that failed, he would no doubt have blown himself up.

The gamble paid off. When dawn broke the next day, *Neptune* was to windward of the Spaniards and sailing away from them. The Spanish put up their helms in a halfhearted attempt to chase, but it was futile. There would be no overtaking the lightened, well-handled ship. The wind filled in from the southeast and *Santo Cristo*'s superstructure, battered in the previous day's fight, collapsed. When *Concepción* hove to to stand by her damaged companion, *Neptune* made a

Naval battle, 1670

clean getaway. It was a remarkable feat for de Graff and a humiliation for the Armada de Barlovento.

Admiral Andrés de Ochoa died a few days after the battle. By the end of September, the Armada de Barlovento had returned to Vera Cruz. For the admiral, death at sea was undoubtedly preferable to enduring the further humiliation of the court-martial that followed that debacle, at which most of the officers were found guilty of misconduct.[1]

The armada did have one thing to show for its efforts: Pierre Bot and his men. Reneging on their promise of quarter, the Spanish tried them for piracy and found them guilty. Bot and his officers were executed, along with six Spanish subjects who had joined Bot's crew. As was usual in such cases, the remaining prisoners were probably enslaved.

De Graff made his way to Cuba. The close call had not dampened his enthusiasm for making war on Spain.

THE LAST OF DE GRAFF, PIRATE

The sack of Campeche had been the Chevalier de Grammont's piratical swan song. De Graff had more left in him.

In February 1686, the Spanish staged a raid on French Saint-Domingue on the island of Hispaniola. De Graff was now so prominent a citizen that he owned a plantation on that island. The Spaniards raided his plantation and carried off one hundred of de Graff's slaves. His wife, Marie-Anne, and their two young daughters narrowly escaped capture. In retaliation, de Graff organized another of his trademark raids. This time he gathered together seven ships and more than five hundred filibusters.

Just as de Cussy had been trying to bring de Grammont under government authority, so he was trying to recruit de Graff. In the fall of 1686, de Cussy wrote to Governor Molesworth of Jamaica, "It is uncertain whether [Laurens de Graff] is gone, but certainly my letter offering him terms has never come to his hands."[2] Or perhaps de Graff simply chose to ignore de Cussy's terms, opting instead for one last raid he knew would never receive official sanction.

De Graff sailed for familiar territory. Once again he went to leeward and the Yucatán, anchoring in Bahía de la Ascención, just over one hundred miles south of their old rendezvous of Isla Mujeres. This time, their target was not a port city. Instead, de Graff's object was the city of Tihosuco, sixty miles inland.

Five hundred buccaneers under the mulatto leader marched against the town, but there was no chance of surprise with so big a band traveling so far. The townspeople fled before the pirates arrived, and de Graff and company looted and burned what was left.

From Tihosuco the pirates continued to push inland toward the town of Valladolid, about thirty-five miles north. As was generally the case, a great army of refugees rushed on ahead of the buccaneers, trying to save themselves and their valuables. Soon there were only thirty-six Spanish soldiers left to defend Valladolid. The buccaneers were virtually unopposed. Then, six miles from the city, de Graff ordered his men to turn and go back the way they came.

No one knows why de Graff retreated with the city right before him and essentially undefended, but a fine legend has grown up around the incident. As the story goes, the refugees, fleeing before the pirates, littered the ground with whatever they could no longer carry. The pirates in turn eagerly gathered up the items left behind.

A quick-thinking mulatto named Núñez saw this and planted a set of fake instructions in one pile of cast-offs. The instructions, per the legend, were supposed to be from the local military commander, Luis de Briaga, ordering that the pirates be lured farther inland and into a trap.

If this is a true story, perhaps the mulatto Núñez missed his calling. De Graff had proved that for a person of color in those days the pirate community was the best place for advancement, and a man who could outwit de Graff would have gone far.

De Graff abandoned the Yucatán and made his way to his other frequent hideout, Roatán. From there he sailed back toward Petit Goâve, but had the bad luck to wreck his ship off Cartagena while chasing a fourteen-gun Spanish bark.

De Graff was not the kind to let a simple shipwreck stand in his way. As de Cussy explained to Molesworth, "Laurens was wrecked off Cartagena while in pursuit of a small bark, but nevertheless took her with his boat and saved his people."[3] In October 1686, he sailed into Petit Goâve, aboard the unfortunate bark.

DE GRAFF—A KING'S MAN

There is some question as to when de Graff made the shift from pirate to French officer. Certainly he had for most of his career carried some commission or other from the various French governors in the West Indies, but that alone hardly granted him unquestioned legitimacy. As we have seen, nearly everyone carried some sort of a commission, including de Grammont and the notorious Van Hoorn.

In October 1687, Molesworth reported that de Cussy had informed him that "the French King had made Grammont (whom we took to be lost) his second lieutenant, and Laurens his third major."[4]

It is interesting to note that de Cussy does not seem to be very forthcoming about de Grammont, who had disappeared a year and a half before, and was unwilling to confirm rumors of de Grammont's disappearance. Perhaps the French governor did not want to admit to his English counterpart that so effective a leader as de Grammont had been lost. That being the case, perhaps de Graff really was not on the French payroll at that time, and de Cussy was just engaging in a bit more disinformation. Whatever the case, de Graff's activities were moving more in line with official French policy in the Caribbean.

Under the heading "too little, too late," the Spanish dispatched a squadron of Basque privateers from the Bay of Biscay in 1687 for the express purpose of hunting pirates and other interlopers in the Spanish West Indies. The commander of the squadron promised "to go in search of the pirate Lorencillo before anything else."[5]

In May of that year, a single frigate from that squadron encountered de Graff on the southern coast of Cuba. As the two ships engaged, the Biscayan frigate promptly ran aground. The situation looked bad for the pirate hunter until a small fleet of Cuban coast guard vessels sortied out from the shore in support of the stranded ship. Rather than abandon the fight, however, de Graff turned on the coast guard vessels, inflicted terrible casualties, sank a piragua, and took a small vessel as a prize.

No doubt de Graff had racked up any number of enemies over the years, but the slaughter he inflicted on the Cuban *guarda del costa* produced an unusually determined foe. Blas Miguel, a Cuban filibuster and perhaps himself a mulatto, had lost his brother in the fight with de Graff. Blas Miguel swore revenge.

A sea battle, 1692

An Act of Ill-Conceived Vengeance

During the dark morning hours of August 10, 1687, Blas Miguel stood into the harbor at Petit Goâve with eighty-five men and two small vessels, a brigantine and a piragua. August 10 is the feast day of St. Lawrence, de Graff's patron saint, and Blas Miguel hoped to catch him celebrating and off guard.

The attack began well. The Cubans stormed ashore at first light and caught the town by surprise. Raging through the streets as wildly as Laurens de Graff ever did, they hacked the mayor to death and bayonetted his pregnant wife. They looted a number of homes and took the small fortress without resistance.

In the growing light of day, it became clear to the residents of Petit Goâve that the invaders were few in number. In fact, of the eighty-five men that Miguel had with him, twenty had remained aboard the vessels, and those ashore were not particularly well armed.

Reinforcements poured in from the countryside. Blas Miguel and those of his Cuban raiders who were still alive retreated to the small fortress. In the harbor, his brigantine abruptly departed in a hail of

Breaking with a wheel

French cannon fire. De Graff himself waded out through the surf sword in hand and captured the piragua.

Trapped and surrounded, Blas Miguel made a laughable offer: to return all the booty he had taken if he and his men were allowed to sail away. Instead, the buccaneers of Petit Goâve stormed the fortress and took Miguel and forty-seven of his men.

The next day, forty-eight surviving raiders were tried, with predictable results. Miguel and two of his officers were sentenced to be "broken alive on the wheel." Two were found to have been forced into the Cubans' service, and they were released. The rest were sentenced to hang.

The punishments were carried out the next day. Attacking the buccaneer stronghold of Laurens de Graff with sixty-five lightly armed men had not been a prudent move.

35

A Naval Officer at War

*Laurens with a ship and 200 men touched at Montego Bay the
other day and did no harm. . . .*
—*Sir Francis Watson*

SUMMER 1687
ILE À VACHE

By summer 1687, de Graff was no longer an outlaw, taking ships
and sacking towns on his own authority, but rather was taking
his orders from de Cussy. In September, the French governor dispatched him to Ile à Vache, under the guise of reinforcing the
French claim to that island off the southern coast of Hispaniola. In
fact, an old Spanish shipwreck had recently been uncovered there
and de Graff's presence was meant to discourage any but the French
from working the wreck.

Laurens was appointed a major, or royal adjutant, of Ile à Vache.
De Cussy reported that the former pirate promised "to acquit himself with the same zeal and fidelity as he had done during the ten
years he has served under the French standard."[1] That was exactly
what he did.

The British were quickly informed that Ile à Vache was now off limits. Molesworth was not concerned, as his intelligence assured him that
the island was of no great value. Not everyone agreed with that. Nearly
a year later, the Duke of Albemarle, who succeeded Molesworth as governor of Jamaica, wrote to the Lords of Trade and Plantations saying that

the Isle of Ash [Ile à Vache], once dependent on Jamaica, and valuable for turtle fishing, has for the past two years been taken by the pirate Laurens, and British subjects have been prohibited from hunting or fishing. The place is of importance, and in case of a war would, in French hands, be very prejudicial to us. . . . [2]

From fall 1687 to fall 1689, Laurens de Graff seems to have lived an unusually peaceful existence, at least by his standards.

After asserting French dominance over Ile à Vache, de Graff received a new mission from Governor de Cussy. Information from a captured Spanish captain alerted them to the presence of a valuable wreck on the Serranilla Bank, a submerged mountain on the ocean floor, nearly equidistant from Jamaica and the east coast of Nicaragua. De Cussy suggested that de Graff work the wreck.

When de Graff sailed for the Serranilla Bank, no one in the Caribbean knew that the nations of Europe were once again at war. This time it was the War of the Grand Alliance, also known as King William's War, pitting France against England and its allies. Even though they did not know that de Graff was officially their enemy, the English kept a wary eye on the filibuster. The acting governor of Jamaica, Sir Francis Watson, wrote, "A number under Laurens have left Petit Guavos [sic] after a wreck, as they give out."[3]

Though Watson does not appear entirely convinced that de Graff was on so benign a mission as fishing an ancient Spanish wreck, that was exactly what he was doing. Using grappling hooks and Indian divers, the filibuster worked the site with limited success for a month or so. When the ship de Graff had dispatched for more supplies failed to return in a reasonable time, he was forced to up-anchor and to search out supplies himself on the southern coast of Cuba.

The former pirate soon learned of the new conflict in Europe, and he knew that he would be called upon to join the fight. He touched at Jamaica, but rather than raid the island, he left a brazen and terrifying promise to return. Watson reported to the Lords of Trade and Plantations:

Laurens with a ship and 200 men touched at Montego Bay the other day and did no harm, but said he would obtain a commission at Petit Guavos and return to plunder the whole of the north side of the Island. The people are so affrighted that they have sent their wives and children to Port Royal.[4]

De Graff was either exhibiting a concern for protocol that had been quite lacking during his days of wanton piracy or perhaps was simply displaying his customary panache.

When Pierre-Paul Tarin de Cussy first took office, he maintained what amounted to a "don't ask, don't tell" policy toward the filibusters, recognizing their importance in the defense and prosperity of the French West Indies. By 1687, in part as a result of the sack of Campeche, the French court ordered de Cussy to reverse that policy and, against his better judgment, he reined in the buccaneers.

As war came upon them, it became clear that de Cussy's initial policy would, in fact, have been the best course. The governor wrote bitterly that if he had been allowed to continue the wink-and-a-nod attitude toward the pirates, "there would be ten or twelve stout ships on this coast, with many brave people aboard to preserve this colony and its commerce."[5] De Cussy understood that maintaining a large fleet of privateers gave the government a navy of sorts at no expense.

Fortunately, de Cussy still had Laurens, now Major Laurens de Graff, Knight of the Order of St. Louis, and de Graff had not forgotten his promise to the people of Jamaica.

DE GRAFF RETURNS

In the beginning of December 1689, de Graff returned to Jamaica, having called at Saint-Domingue to receive official orders from de Cussy and to recruit a small fleet of French filibusters. Despite de Graff's earlier threat, official word of the hostilities in Europe had not reached Jamaica. De Graff was able to scoop up eight or ten unwary English ships as prizes, as well as to stage a raid on at least one coastal plantation.

The English hurriedly assembled a fleet under the command of Captain Edward Spragge of HMS *Drake*. The minutes of the Council of Jamaica reflect the hasty preparations. "On the report of the pirate Laurens, ordered that the Island's armed sloop come to Port Royal to join the fleet against Laurens, and that a second sloop be fitted out. . . . Order for pressing a ship for the fleet against Laurens."[6]

Ship-to-ship

The minutes also reflect the realization that it was no longer safe to venture out to sea with de Graff lurking offshore. Six days later the council ordered "the sloops not ready to accompany Captain Spragge against Laurens, not to leave the harbor."[7] And three days after that, "that the known trading sloops and no others be allowed to leave the harbor after the departure of the fleet against Laurens."[8]

Implicit in this order is the fear that spies might warn Laurens of the preparations being made against him.

The little fleet had no effect against the pirate and his squadron. HMS *Drake* was in such poor shape that she was condemned early the following year, so it is hardly surprising that Spragge could not drive the filibuster away.

In March 1690, a second attempt was made to expel Laurens from the Jamaican coast, with similar results. In fact, it was not until the end of May that he finally sailed from that island. Laurens de Graff had terrorized the coast of Jamaica for half a year and had held it under the thumb of his blockade.

When HMS *Drake* was condemned, the Council of Jamaica was forced to dispatch its armed sloop to the Cayman Islands to protect the English turtling vessels working there. De Graff was also making for the Caymans. The Earl of Inchiquin, who was then governor of Jamaica, reported the incident to the Lords of Trade and Plantations:

> The Island has therefore fitted out a sloop, which lately went to Caymanos for turtle, where there were several of our craft lying. There Laurens, the great pirate of Petit Guavos, engaged the sloop, and the rest of the craft escaped. The firing was heard continuing till eleven at night, and as this was a month since and nothing has been heard of the sloop, we conclude that Laurens has taken her, he having two men against one in his barco longo. We have therefore no ships now except HMS *Swan,* which is so bad a sailor that she is little better than nothing.[9]

It was one of the few occasions in which a pirate was able to defeat an English naval vessel. The sloop apparently put up a terrific fight before she was taken, fighting well into the night. Inchiquin's complaint concerning a lack of ships reflects the ongoing problem of European nations' unwillingness to dispatch their best ships to the West Indies and thus strip the home defenses. The naval forces in the Caribbean and the American colonies tended to be older, second-line ships.

Thomas Lynch, the former governor of Jamaica, had carried on an amiable correspondence with de Graff, with the hope, of course, of luring him into the English fold. Now that de Graff was fully on the side of the French, the new governor had quite a different attitude. No matter what commission de Graff held or what title was given him by the king of France, he was always labeled a pirate by the English.

After taking the English sloop in the Cayman Islands, de Graff returned to Saint-Domingue, having heard rumors of a joint English-Spanish raid on the island. De Cussy transferred de Graff's base of operations from Ile à Vache to Cap François, as he did not wish to "risk further a person so zealous in his service in such a feeble quarter."[10]

COMBINED OPERATIONS

In January 1691, de Cussy was ready to lead an offensive against the Spanish near Santo Domingo. With de Graff's filibuster fleet for transportation, he landed with a small army at Saint-Domingue and marched inland. Past Spanish military performance in the region gave de Cussy and de Graff cause for optimism in their venture.

This time, they met with something that de Graff had never seen: fierce and overwhelming Spanish force. The Armada de Barlovento landed 2,600 Spanish troops near Cap François. Another seven hundred made the march overland from Santo Domingo. They outnumbered the French three to one.

Such odds had never been a problem for de Graff in the past. He had always come out on top. This time, however, the Spanish soldiers were not poorly trained and unmotivated militia or garrison troops, but well-disciplined and well-prepared line infantry.

French and Spanish met on an open plain called Sabane de la Limonade, and what began as a battle soon became a rout. The Spanish killed as many as five hundred of the French troops, including Governor Pierre-Paul Tarin de Cussy.

Laurens de Graff was the man the Spanish wanted most, but he managed to flee into the hills. In the weeks while the Spanish rampaged throughout the countryside before finally withdrawing, de Graff narrowly avoided capture. Arrogance and an underestimation of the fighting capability of the Spanish had resulted in the loss of the governor's life and the filibuster's greatest defeat yet. And there was more to come.

Governorship of Saint-Domingue passed to Jean-Baptiste Ducasse, an energetic and able man. With the loss of so many officers in the disastrous battle at Sabane de la Limonade, de Graff now assumed an even more important role in French military activity in the West Indies. At this point, the former slave became Sieur de Graff, lieutenant du roi for the government of Ile la Tortue and coast of Saint-Domingue.

In 1692, de Graff was busy recruiting, organizing, and maneuvering French forces against another possible Spanish invasion, but the Spanish never came. Despite de Graff's three years of land fighting, Ducasse recognized that the filibuster would be far more effective afloat than he was leading ground troops. Having de Graff loose at sea would help keep the enemy off balance. By 1694, the governor was

A buccaneering battle

ready to employ de Graff in the manner in which he had been so effective for the past decade: leading a massive buccaneer army in an amphibious raid.

For the first time in his career, de Graff would be leading his filibuster army against the English, not the Spanish. The target was Jamaica. In June 1693, de Graff and Ducasse organized an armada consisting of twenty-two ships and more than three thousand filibusters. Among these men were English and Irish Jacobites, supporters of the recently deposed James II as the true king of England. With this powerful force, de Graff fell on the eastern tip of Jamaica.

This attack had not even a tinge of piracy. De Graff was a legitimate military officer, a knight of the Order of St. Louis, leading French troops. The sanction of legal authority did not make the former pirate any less effective. He and his men landed at Cow Bay and Point Morant and ravaged the eastern part of the island, then made a feint toward the capital of Jamaica, Port Royal. When the English sent columns of troops to meet them, the seaborne raiders returned to their ships and stood out to sea.

They reappeared on the night of July 28, landing fifteen hundred men at Carlisle Bay. The next day de Graff and his men advanced against a garrison of 250 men. Boldly holding their fire until the last moment, the filibusters were able to deliver a devastating volley that drove the English defenders from their trenches and sent them fleeing.

Unfortunately for the buccaneers, Jamaica is considerably smaller than the Yucatán, towns were not so isolated, and, since the island was the capital of the English West Indies, there also were significant forces stationed there. Reinforcements were sent from Port Royal and arrived after a hard forced overnight march. Only their arrival prevented de Graff from laying waste to the entire area. For nearly a week he maintained control of the ground he had taken, sending his filibusters out to scour the countryside.

The Jamaican plantation houses, however, were each built like mini-fortresses, and the filibusters had no artillery to take them on. They satisfied themselves with whatever loot they could find, as well as nearly sixteen hundred slaves. On August 3, 1694, de Graff and his men reembarked and left Jamaica behind.

It was the last time De Graff led a seaborne raid at the head of a buccaneer army.

The following spring found Laurens de Graff again in command of a small land-based force near his new home at Cap François on the north shore of modern Haiti. On May 24, 1695, the enemy came.

Not just a raid but a major combined operation landed near de Graff's plantation. The British Commodore Robert Wilmot and Colonel Luke Lillingston had arrived in the West Indies and formed a joint operation with their opposite numbers in the Spanish naval and land forces. They completely overwhelmed de Graff and his small army, forcing the French to retreat in the face of their onslaught. De Graff abandoned his home to the invaders. His wife, Marie-Anne Dieu-le-Veut, and their two daughters were captured and made prisoners.

De Graff sent word to Governor Ducasse calling for reinforcements, but they did not come, probably because Ducasse had none to send. The combined English and Spanish troops pushed the French defenders back as far as Port-de-Paix, which the invaders overran more than a month after landing. Having taken Port-de-Paix, the English and Spanish withdrew, leaving behind the kind of destruction that de Graff had once brought to the Spanish.

Once the fighting was over, the finger-pointing began. De Graff bore the brunt of it, with some even suggesting that he had colluded with the enemy, since Holland, de Graff's native land, was allied with the English and Spanish. Laurens was relieved of his command and sent to France to face a court-martial.

The subsequent trial of the Sieur de Graff, lieutenant du roi, completely exonerated him from any wrongdoing. He returned to his home in the West Indies, but by then the war was over, and his standing was greatly diminished, despite having been cleared by the court-martial. De Graff's wife was held captive by the Spanish until the very last prisoner exchange in October 1698, possibly out of Spanish vindictiveness toward its former tormentor.

Laurens de Graff's fighting days were over. By the time the famed buccaneer returned to Saint-Domingue from his court-martial, it had been twenty-one years since the great wreck at Las Aves. Twenty-one years of near constant warfare, of pirate raids and bloody land battles. No doubt, de Graff was weary and ready for a change.

An Old Pirate Moves On

A the end of the year 1698, an explorer named Pierre Le Moyne d'Iberville stopped off at Saint-Domingue. He was en route to Louisiana with an eye toward establishing a colony there. De Graff, now around fifty years old, agreed to accompany him as translator and guide.

At first, de Graff's reputation proved to be a hindrance to the expedition. The five ships of d'Iberville's squadron anchored off of Pensacola, Florida, in January 1699 and called for a pilot to bring them into the harbor. The Spanish officer in charge of the local garrison there was not happy to see ships belonging to his recent enemy just offshore. When he went aboard he was even less happy to find that d'Iberville's translator was none other than the famed "Lorencillo." Entrance to the harbor was refused, and a few days later, at the request of the Spanish, the squadron put to sea again.

The French squadron sailed west, eventually landing at what is now Biloxi, Mississippi. With them went de Graff, who settled there with the official function of clerk for the king. Five years later, it was reported that he was dead.

Just as de Graff's beginnings were shrouded in mystery, so too was his end. One source states he died near Biloxi, but another says he is buried at Axis, Alabama, a suburb of Mobile. The governor of Cap François (modern Cap Haitien, Haiti) wrote that he had died there on May 25, 1704.

The end of the seventeenth century saw the end of the great buccaneers of the Caribbean. The piracy that pitted English and French against the Spanish Empire in the New World was over, and the filibusters of Tortuga and Petit Goâve, who were looked upon by colonial governors as both a plague and a resource, were gone. Some retired, most were dead. The Carribean had no place for them anymore.

The piracy that would spring up to take its place, piracy based from Madagascar and the Bahamas and colonial America, would be something very different.

The greatest of the true buccaneers, Laurens de Graff, died peace-

The pirate's dream

fully in a quiet backwater colony. In his tumultuous lifetime he had gone from slave to knight of the realm, a black man who rose higher than any other filibuster ever did, in a world in which slavery was an integral and unquestioned way of life. Of all the pirates of the Spanish Main, de Graff was the best of them. The world has never since seen his like. Nor is it likely to.

36

From Good to Bad
to Ugly

At first, I was not sure what to do about the pirate wrecks for fear word might leak out to the wrong people. News that we had found a pirate shipwreck with intact metal-laden barrels would bring swarms of treasure hunters.

It should be remembered, of course, that d'Estrées' fleet was on its way to sack Curaçao, not on its way back. There would have been no captured booty on board, and filibusters seldom carried their personal fortunes with them. Still, it conjured up the kind of images to tempt what Billy Bones in *Treasure Island* called "lubbers as couldn't keep what they got, and want to nail what is another's."

After a few hours, I did tell the others of our discovery. They were as excited as I was, and I told the story of de Grammont and de Graff, and how they led the buccaneers of Las Aves on their rampage at Maracaibo.

The next day we went back with the whole team to map and film the site of the second pirate shipwreck. The wind was building, but it was not so bad as to keep our dive boat from getting over the reef. We knew, however, that it was only a matter of time before we would be blown off the site.

We found even more artifacts, including cannons and some odd, concreted shapes, the identity of which we could not even guess at. We also found clear glass liquor bottles embedded in the coral. That was the only wreck where we saw that. It was another identifier: I would expect to see more gin or rum bottles aboard a pirate vessel than a man-of-war. We swept the barrels again, and again the metal detectors sang. Most likely they held gun parts or nails, but one couldn't help thinking, Silver? . . .

The barrels were camouflaged with coral and a fantastic array of other sea growth. They had become a part of the living reef, and to remove them would have taken a lot of brute force that would have irreparably damaged the reef. Even if our permit would have allowed for excavation and retrieval, there would be the question of whether or not to leave them in situ.

As at all of the other wreck sites, there were also cannon balls and lead shot scattered everywhere. We couldn't fan our hands over the sand without seeing them. The shot were in little pockets where the canvas bags that once held them had rotted away. Stored in shot lockers, big wooden bins belowdecks, the cannon balls had scattered all over the reef when the ships broke up. Using the metal detectors became almost pointless because they were going off constantly.

Inspecting a filibuster wreck site

Diver over a wreck camouflaged by coral

I was still thrilled about the wreck, even as we went through our ordinary mapping routine the next day. We had found two pirate-ship wrecks and used an old map to do it. It doesn't get much better than that for pirate hunters.

I kept thinking about how Ken Kinkor was going to react to these discoveries. With the *Whydah,* he was studying artifacts from the first pirate shipwreck ever authenticated. If this project moved from "reconnaissance and survey" to "archaeological excavation and recovery," there would be data from the second and third to be scrutinized and interpreted as well. While these vessels might never be identified by name, they could still serve as a potentially valuable pool of information about a subculture of men whose lives are seriously misunderstood by both scholars and the general public.

From the very onset of my research on the *Whydah* nearly two decades ago, I had noticed the international character of her crew. Ken had taken these perceptions and built further on them. We learned that pirates of the early eighteenth century had practiced an amazingly high level of tolerance among themselves: national, religious, and racial. When compared with European societies and their colonies, buccaneers were also extraordinarily democratic and egalitarian. Now we had the chance to see if their forebears by three decades had held similar beliefs.

Analysis of shipwreck artifacts produces a sense of the men who used those objects. We believe that Sam Bellamy and his crew of the period 1715–25 held most of the same beliefs as the Brethren of the Coast of the period 1678–1700. However, those buccaneers operating prior to the disaster at Las Aves were more closely tied to their respective governments than were the freebooters who followed in their wake. The later freebooters and pirates were far more self-reliant, which led to a different form of social organization. In practical terms, Laurens de Graff would have found himself at home aboard the *Why-dah* in ways that Henry Morgan would not have. Without the disaster at Las Aves, which severely weakened French naval power, the freebooters and pirates of the post-1678 period would not have risen to power.

Having read the history of d'Estrées' fleet and seen the wrecks, I had a good sense for the scale of the disaster in terms of the human cost. And I had already fought my own battle with the reef. In full daylight, knowing exactly what I was up against, with cutting-edge equipment and a lifetime of underwater experience, I had still been nearly knocked out of the ring by the reef.

I thought of those men, the French soldiers and sailors and the buccaneers, who in one instant went from the solid deck of a sturdy man-of-war into the cauldron, half-drowned, dashed against staghorn and fire coral, dragged over the reef by crashing waves in the dead of night. Most people in the seventeenth century could not swim. Swimming would not have done them much good anyway, not on that reef.

No doubt some managed to cling to wreckage and drift to the beach. Others must have walked through the rushing waters across the reef to shore—though the trek would have cut their feet to pieces. In some places they would have had to swim across gaps in the coral. In some instances, the ordeal might have stretched almost four miles. I am amazed that anyone got to shore alive. Many of them, of course, were so cut up by the coral that they wished they were dead. It appears that about half of them were granted their wish.

As they reached the sandy beach at Las Aves, they must have collapsed in exhaustion, relieved to be on dry land. Yet many would have been in agony from the beating they had taken in the turbulent water on the reef and their wounds from the fire coral.

Each time I looked at the island I felt something of the despair these sailors must have experienced. It is one of the most inhospitable, windswept, arid pieces of real estate I have ever seen in the Caribbean. There is virtually no vegetation aside from scrub brush and mangrove, and no shelter from the welding-rod bright sun.

I can only imagine what those castaways thought. Some of them probably gave up hope and died as soon as the rising sun revealed what a godforsaken place they had ended up on.

The sores and lacerations on the sailors' hands, knees, and feet would have started growing septic, hour by hour, day by day. Fire coral leaves a painful red welt that feels as if you are being jabbed with hot matches. The bugs on the island are intolerable. When we were anchored in the lagoon with the breeze blowing we didn't notice them, but, once ashore, they came swarming—flies, no-see-ums, and sand fleas—just as they must have come swarming to the smell of fresh blood from the survivors' wounds more than three centuries ago.

The sun would have beaten down on them and their thirst would have quickly become overpowering. Wine and brandy were their staples, but alcohol would have further dehydrated them, making the agony worse. For food, they had either salt-meat fished from casks drifting in from the ships or conch, which is also very salty. They would have required enormous amounts of water to avoid complete dehydration, and so they must have been going mad with thirst.

I spent some time walking around the shore, thinking of those men cast up on the beach. I looked for evidence of graves but did not find any. The French would have buried their dead properly, perhaps a mass grave for the common soldiers and seamen and individual graves for the officers and gentlemen.

I did find some old foundations made out of conch shells and crude brick hearths. They were ancient structures, most likely long deserted when the buccaneers wrecked on the reefs. No doubt they were already no more than foundations when de Grammont was cast up on the beach. These had been European-style houses, the biggest perhaps ten feet by twenty. They represented an earlier, unrecorded attempt to settle there. Nothing in the record indicates that these houses were still habitable at the time of the wrecks on Las Aves. They had belonged to settlers who had given up.

I pictured filibusters sheltering themselves from the wind, decked out in finery they had salvaged from the surf, drinking themselves into happy oblivion with wine and brandy liberated from the holds of the men-of-war.

Ancient cellar on Las Aves

The freebooters were hard men. Amid all that death and suffering, a tragedy of epic proportions, legend says they were toasting the short and happy life of the buccaneer. Eat, drink, and be merry, for tomorrow we die. It is hard to imagine people so fatalistic and careless of suffering—their own or anyone else's.

37

Cannons, Anchors, and Surf

On our sixth day of work, just as the conditions were going from bad to worse, we made an extraordinary find: a pile of iron cannons, fifty or more of them, and an anchor, all heaped together. It was unlike anything I had ever seen. When a ship goes down as they did when the fleet struck Las Aves, the cannons tend to be scattered around the seabed, or fall off as the ship rots away. But not here. These were piled up like logs in a logjam. It was not a random accident.

Someone had done this, and from the coral growth, done it a long time ago. It might have been Thomas Paine in his effort to salvage what he could find. It might have been d'Estrées himself, who did return to the island some months after the disaster to recover some of his lost fleet, and perhaps some of his reputation. It could also have been any of the thousands of buccaneers, wreckers, and human flotsam who prowled the Caribbean world, aware of the riches that lay strewn over the reef.

I asked Chris Macort to make a careful drawing of the site. We all wondered, why did this happen? Why were the cannons deliberately piled in that way?

From the size of them, they looked to be from one of the larger ships, maybe *Le Terrible*. Once the ship hit the reef, the crew might

Barry Clifford and the pile of fifty cannons

have lowered the cannons overboard with a block and tackle to lighten the ship in an attempt to get her off. Or it might have been part of a salvage effort. Iron cannons become useless as artillery after they have been submerged in seawater for any length of time. Bronze guns can be salvaged, however. Perhaps the salvagers pulled the guns up one at a time, kept the bronze ones, and threw back any iron guns.

Just past the halfway point in the expedition, the wind was back with a vengeance, blowing nearly as hard as it had on our first trip. The surf was exploding against the reefs with a force like an earthquake, sending shock waves through the water that could be felt a quarter mile away. We no longer had the luxury of motoring over calm water to look for wrecks. We once again had to pull ourselves against the current over the shallow reef.

It was hard and brittle. We tried to be as gentle as possible as we pulled ourselves over the reef, but it was hard not to break the delicate coral. It was much like cave diving, with its delicate stalactites and stalagmites. I realized the disastrous effects dozens of divers might have over the course of many years.

There was more than just coral, of course. The reef is alive with a wide range of undersea flora and fauna: sea fans, gorgonias, brain coral, staghorn coral, and all of the other creatures dependent on that ecosystem. It was beautiful, but that beauty was hard to appreciate when we had to drag ourselves over it just to get to the wrecks.

Once again, the conditions were just like being inside a washing machine. We would stick to the bottom, which was only a few feet down, weaving our way through the taller stands of coral. When the big waves we called gorillas came, we would blend into the bottom like flatfish and wait for it. We wore leather gloves for just that purpose. The waves would crash on top of us, and we wouldn't see a thing except froth, bubbles, and turbulence.

As the water rushed out, we'd kick like hell until the next one hit, and then we'd stop and hang on again. It was hand over hand over hand and stopping while the waves battered us. You couldn't stand up in it—you'd get knocked over or get your legs knocked out from underneath you. So we clawed our way along, burning air, exhausted and frustrated that we were reduced to this once again.

As on our first trip to Las Aves, coming back in over the reef was a wild ride. Every day the wind blew stronger, and the danger increased proportionally.

On the ocean side of the reef, it was ten or twelve feet deep in this area. There were a couple of cannons in three feet of water, but about 80 percent were in ten to twenty-five feet of water, which made working somewhat easier.

Once we took the *Aquana* outside the lagoon. With the big surf breaking, the only way to get the dive boat out was to motor all the way around the tip of the reef, several miles from the *Antares*. It was almost as dangerous, however, as crawling over the reef.

What we really wanted to do was to move the *Antares* to the edge of the reef to make the trip in the dive boat shorter. That would have put us in plain sight of the coast guard station and the navy ship, like a mouse making its nest in plain view of the cat's favorite chair.

The coast guard or the navy continued to board us on a regular basis, and we'd go through the same routine. We were actually getting to know them. Many of the Venezuelans asked to have their pictures taken with us. We were happy to oblige.

Mike's people were still working to get the navy to recognize the validity of our permits. They were making headway, but, as in dealing with any military organization, it can take a while for decisions to work their way down through the chain of command. We were on a tight schedule. We knew we had valid permits, but there was no sense in flaunting what we were doing until the local commander had received his instructions from Caracas. So we left the *Antares* where it was and either swam over the reef or took the long ride around the outside.

The big seas made it difficult not only to get to the wrecks, but to work them once we were there. During the first part of the expedition the divers had the luxury of freely swimming around the wreck sites in relative safety. Those days were gone.

Now the current was running strong, even in the deeper water outside, threatening to sweep divers into the reef zone. We needed to maintain handholds as we worked or to keep kicking like hell. The surge from the big waves created aftershocks that bounced us around like rag dolls in the mouth of a bull terrier. If we let go of whatever we were holding on to, we would get swept away and smashed up against coral like the surface of a cheese shredder.

The current would rip things from your hands. The hundred-foot tape measure that Carl and Todd were using became a nightmare as the current carried it off, twisting and tangling it around coral and artifacts. Chris's underwater notebooks were wrenched from his grip. The work was much tougher now, and much more dangerous. If it got any worse, our expedition would be cut short.

The next day, November 3, we moved north along the reef. The big freighter stuck high on the reef had gone aground right at the location of one of the larger wrecks, perhaps d'Estrées' flagship, *Le Terrible*. In fact, the modern vessel probably went aground at the very point where several of the bigger French ships had struck. The freighter captain could not have found a worse place from an archaeological point of view. I imagine, however, that he had little choice in the matter.

The freighter had done massive damage to the reef and the wreck site, and added some drama of its own. On the bottom, we found a navy stockless anchor, the standard anchor on a modern steel ship. Still attached to the ring, a great length of chain snaked across the reef. Draped over staghorn and fire coral and the debris field of d'Estrées' flagship, it led right to the wrecked freighter from which it came. This was a vivid reminder of the freighter crew's last desperate attempt to keep their ship from the killer reefs, like those two ancient anchors hooked together. Their effort, like the Frenchmen's before them, was in vain—the modern version of d'Estrées' nightmare.

I don't know how the freighter came to be wrecked or if any of the crew were killed. I doubt very much that they were. I doubt anyone

Coasting freighter wrecked on the reefs
of Las Aves

was even hurt. The ship held together, save for the big gash in her bottom. Steel hulls do much better on reefs than wood. In this case, it was the reef that got the worst of it.

But the sea is as patient as Ah Puch, the Mayan god of death, and will win in the end. For all her modern materials and construction, the freighter is quickly oxidizing away. Like the French fleet, someday she will become an indistinguishable part of the reef.

She was also making it difficult to explore and map the site. We were all afraid that the swirling current around her would sweep us through the jagged, rusty hole in the freighter's side, uglier and more intimidating even than a barracuda's mouth. We kept a sure grip on cannons and anchors as we worked upcurrent from that threat.

There was no way to know just how much of the old French wrecks had been disturbed by the passing of the freighter over the reef, but a wide swath was mowed through the coral and the debris field. We found twisted, rusting I-beams lying on top of seventeenth-century cannons.

In fact, there were even some cannons underneath the wrecked freighter. It was surprising that we could see them at all, but there was a hollow space under the wreck, a narrow cave maybe four or five feet deep with the massive, rusting hull forming the roof. Into that dark, narrow space Chris went with his tape measure, to make sure we recorded every cannon possible.

The bottom was littered with all of the detritus dumped out of the split bottom of a massive warship. If we had been digging, we could have spent months just excavating that single site. But that was pure fantasy. Now, with conditions deteriorating, we were only hoping to complete the limited mission we had set for ourselves, and to get out of there before something bad happened.

38

A Pirate Reaches Retirement

Perhaps you will consider whether our ambassador should not procure the French King's orders on the subject [of suppressing piracy], for saying anything here is like preaching in the desert.
—Sir Thomas Lynch to the Lord President of the Council

SUMMER 1690
RHODE ISLAND

The great age of the buccaneers, the de Grammonts, the de Graffs, who plundered the Spanish Main was coming to an end. This is not to say that the curtain was falling on piracy in general. Far from it. The golden age of piracy would continue on for another forty years before piracy would dwindle to the point at which it was a rarity, an anomaly, in the history of merchant sail.

With the filibusters of old gone, the pirates of the Caribbean would rise in their place. There was, of course, a continuity, a thread that ran from the history of the old to the dawn of the new. There were men who spanned that historical gap. One of the most important was de Grammont's old lieutenant, Thomas Paine.

Paine was an odd figure, a shadow pirate, as often operating in the background as he was leading raids. He outlasted de Grammont, de Graff, and most of the others. He was a major link between the European buccaneers of the Spanish Main and the pirates that later swarmed out of New England for the Red Sea, and later still the second wave of pirates of the Caribbean, following the War of the Spanish Succession. Paine was one of the few who lived to old age and genuine respectability in the colony.

After the legal close calls and active attempts to prosecute him in 1683 and 1684, Thomas Paine seems to have disappeared for a while. He might have been lying low, waiting for the storm to blow itself out. He could have returned to the Caribbean for another go at filibustering. Although there is no direct evidence that he was there, his old consort Bréha was active in the area, along with Yankey Willems. A couple of English sloops were reported to have been operating in company with those two notorious buccaneers. One of those might well have been commanded by Paine.

Whatever Paine was doing, it does not seem to have further damaged his reputation. By 1687, he turned up again in Rhode Island with little fanfare. There were no attempts to arrest him. That year or the next, Paine married the daughter of a prominent Rhode Island citizen, Caleb Carr, a judge living in Jamestown.

Paine and his bride, Mercy, also settled in Jamestown. Though he had not yet been made a freeman of the colony, that is, an enfranchised voter, Paine served on the grand jury in December 1688. He was becoming a respectable citizen. The government was now willing to overlook his past indiscretions.

A Pirate for Rhode Island's Defense

One reason for governmental tolerance of pirates was the hope that these men would form a floating militia in times of crisis. In 1690, Rhode Island would have reason to be glad the former buccaneer was in their midst.

The War of the Grand Alliance, or "King William's War" as it was known in the colonies, had been going on for two years. The governors of New York and Massachusetts had already organized attacks against French colonies in North America. The towns along the seaboard were braced for a counterattack.

On July 22, 1690, a squadron of vessels, a bark and two sloops, appeared off Block Island. Immediately alarmed, the islanders hurried to the shore to determine the identity of the approaching ships. A boat put off from one of the strange vessels, and a man identifying himself as William Trimming, or possibly Tremayne, came ashore. To the colonists' vast relief, he assured them that the ships were English privateers.

Unfortunately, he was lying, but the people of Block Island bought the story. Soon after, when more boats from the squadron came ashore, the people of Block Island made no effort to resist. They did not even realize they had been tricked until the boat crews snatched up hidden weapons and leaped ashore, taking many of the islanders prisoner and sending others fleeing into the woods.

Trimming turned out to be captain of one of the sloops, described by a witness as "a very violent, resolute fellow,"[1] but he was not the leader of the expedition. That honor went to Pierre le Picard. Picard, like Paine, was a filibuster of long standing, one of the old-guard privateers of the Spanish Main.

Pierre le Picard first appears in the history of the filibusters in 1668, twenty-two years prior to the Block Island raid, an incredibly long career given the inherent dangers of the business. The existing record suggests that he first sailed under the command of another pirate, the Frenchman Francis L'Ollonais.

Though Picard seems to have begun his piratical career under the French madman, there is no evidence to suggest that he ever engaged in cruelty of L'Ollonais's caliber. Picard and his men took Block Island with virtually no resistance and occupied it for a week. Contemporary reports say that the privateers brutalized and maltreated the inhabitants, but that maltreatment did not rise to the level of what Exquemelin attributes to L'Ollonais.

When word of the privateers' presence reached the mainland, bonfires were lit up and down the coast as a warning to ships and coastal residents. A sloop was dispatched from Newport to determine the Frenchmen's whereabouts. On July 24, Picard and his men abandoned Block Island and made an attempt on Newport itself, but abandoned that plan when their intentions were discovered and the citizens of Rhode Island forewarned.

To drive the French off by force of arms, Rhode Island governor John Easton commandeered a sloop of ten guns, the *Loyal Stede* of Barbados, then at anchor in Newport Harbor. To command her he chose the Rhode Island citizen most experienced with naval warfare, Captain Thomas Paine.

Pierre le Picard and Thomas Paine had both been Brethren of the

Coast. They certainly knew each other from the old days on the Spanish Main. There is even evidence to suggest that Picard had once served under Paine's command. Now, because of an outbreak of war two thousand miles away, they were bound to fight one another.

Paine set sail aboard the *Loyal Stede* on July 30 with sixty men aboard, including his father-in-law, Caleb Carr, and two brothers-in-law, Nicholas and Samuel Carr. In company with the *Loyal Stede* was a smaller sloop under the command of John Godfrey. Thomas Paine was finally going out to "seize, kill, and destroy pirates," seven years after Sir Thomas Lynch had issued him a commission to do so.

The two improvised men-of-war sailed for Block Island but found it abandoned by the Frenchmen, who had sailed off to try a quick raid on New London, Connecticut. The next day "Captain and Commodore Paine" got his small squadron under way, heading southward, and later in the day he caught sight of the French privateers sailing to the eastward. The Frenchmen, thinking the two sloops to be merchantmen and possibly valuable prizes, hauled their wind and came after them.

Paine was outmanned and outgunned, and he knew he was not dealing with a timid opponent. He brought his sloops into the shallows near Block Island and anchored them fore and aft so they would not swing. In that way Paine kept his broadsides trained on the approaching vessels and prevented the French from getting to either side of his own sloops.

Picard still did not know with whom he was dealing. He still believed that the two sloops were unarmed merchantmen. Rather than go after so easy a prize with his larger vessels, the Frenchman filled a piragua with armed men and sent her in after the anchored sloops thinking a few volleys of small-arms fire would induce the merchantman to surrender. And it probably would have, had the sloops been merchantmen.

In a sort of Bunker Hill don't-fire-until-you-see-the-whites-of-their-eyes strategy Paine and his men waited patiently while the piragua closed with them. Unfortunately, Paine's gunner was overly eager to have at them. He urged Paine to let him fire, arguing that he could put a cannon ball right down the length of the piragua, from bow to stern, doing terrible damage.

Paine argued that they were better off letting the enemy get closer still. With the kind of town hall atmosphere that could only exist among an ad hoc crew of Rhode Islanders, the gunner persisted in his arguments. Finally Paine allowed him his shot, since the gunner was "certain (as he said) he should rake them fore and aft."

The gunner fired and missed, and Picard realized that he was not dealing with an unarmed merchantman. The piragua turned on its heel and headed back to the large privateers, robbing Paine of his chance to dispose of a good portion of the enemy's crew with a single broadside.

The men in the piragua put their backs into their oars to flee the *Loyal Stede*'s broadsides. Once out of range, they waited for the squadron of privateers to sail up to them and then reembarked. The battle would now be ship to ship.

Picard, in command of the bark, led the attack, as the three French vessels approached in line-ahead formation. The bark swept down on the anchored Rhode Island vessels. Coming up with them, the three ships exchanged murderous broadsides of great guns and small-arms fire as Picard sailed slowly past.

Next in line was Captain Trimming, in command of the larger of the two sloops. He went about his business like a true pirate and died the same way. Eyewitnesses reported, "He took a glass of wine to drink, and wished it might be his damnation if he did not board them [the Rhode Island sloops] immediately. But as he was drinking, a bullet struck him in his neck, with which he instantly fell down dead. . . ."

The death of Trimming did nothing to quell the fury of the battle. The fighting was fast and furious. Samuel Niles, an eyewitness, reported:

> [T]he large sloop proceeded, as the former vessel [Picard's bark] had done, and the lesser sloop likewise. Thus they passed by in course, and then tacked and brought their other broadsides to bear. In this manner they continued the fight until the night came on and prevented their farther conflict. Our men valiantly paid them back in their own coin, and bravely repulsed them, and killed several of them.

The battle lasted for several bloody hours, but the English gunnery proved superior to the French. Picard lost fourteen men killed, including Trimming, which he considered a serious blow. He was reported to have claimed that he would rather have lost thirty men than his valuable second in command.

In comparison, Paine lost only one man killed, a Native American, and six were slightly wounded. The French had aimed high, and a majority of their shot passed over the anchored sloops, so that the thrifty Yankees were able to collect musket and cannon balls on the shore beyond Paine's fleet.

As night fell, the French moved offshore and anchored for the night, not far from Paine. The English had nearly exhausted their powder and shot during the long engagement. Paine, expecting the battle to resume at first light, sent for whatever supplies might be found on Block Island.

Pierre le Picard, however, had no more stomach for the fight. Rather than engage the English again, his squadron weighed their anchors at dawn and sailed off, heading out to sea. Certainly the results of the previous day's fight would have been enough to discourage him, but Niles offers another possibility:

> [O]ne reason might be this (as was reported) that their Commodore understood by some means that it was Captain Paine he had encountered, said, "He would as soon choose to fight the devil as with him." Such was their dialect.

For once, Paine's unsavory reputation did him some good.

Seeing the Frenchmen making their escape, Paine and Godfrey went in pursuit of them, "with the valor and spirit of true Englishmen," according to Niles. The privateers were fast and weatherly ships, however, and Paine could not overtake them.

Picard also had in his company a small prize he had captured during his attacks on Long Island Sound, a merchantman loaded with wine and brandy. This ship was a dull sailor compared to the privateers, and Picard knew that she would not be able to outsail Paine. Rather than let the English recapture her, the French blew a hole in her bottom with a cannon and allowed her to sink. It must have been heartbreaking for a French crew to see a cargo of wine and brandy go to the bottom.

When Paine reached the scuttled merchant ship, he found her hanging in the water. The bow had settled onto the bottom, but her stern was still held above water by a line made fast to a longboat that the merchantman had been towing astern. With the ship in that odd situation there was no way to salvage any of the cargo. When the Englishmen cut the line, the ship sank immediately. The Frenchmen were heading for the horizon with no chance that they would be overtaken. Paine and Godfrey had only the longboat as a prize.

But prizes were not the issue here. Commodore Paine had attacked and beaten off a superior enemy, an enemy that had already caused tremendous harm to the coast and threatened to cause even more. Paine the "archpirate" was now Paine the hero, the savior of Rhode Island.

THE GOLDEN YEARS

During the next decade, Thomas Paine continued to grow in wealth and respectability. Only two months after driving off Picard and his squadron, Paine and his father-in-law Caleb Carr were made tax assessors for their hometown of Jamestown.

Two years after that, in 1692, Paine was appointed by the general assembly to the rank of captain of militia. Generally such appointments are made by the town from which the militia is mustered. In this case Jamestown had failed to do so, letting the job fall to the assembly. The fact that it was the colonial government, not the town government, that selected him is evidence that Paine's reputation was not only good but also widely known, even at the level of the colony's general assembly.

Paine's reputation and notice were much enhanced in 1695 when Caleb Carr was elected governor of Rhode Island. Rhode Island politics was (and still is) raucous and colorful. Nepotism was an art form. Even by the standards of the seventeenth century, when nepotism was not nearly as frowned upon as it is today, Rhode Islanders were notorious for the practice. Having a father-in-law as governor could not have hurt Paine's community standing.

In 1698, Thomas Paine was officially admitted as a freeman of the Colony of Rhode Island and Providence Plantations. What is most surprising is that he had not been admitted earlier. Many of the posts he had already filled—tax assessor, militia officer, member of the grand jury—were generally reserved for freemen. His final admission as a freeman seems to have been a mere formality.

Paine had finally arrived as a member in good standing in Rhode Island society. He was washed clean of the stain of piracy and its attendant scandal. And save for one uncomfortable incident in 1699, when he was accused, with good reason, of hiding loot for the notorious Captain William Kidd, he remained an upstanding citizen—at least by Rhode Island standards.

In 1701, when Kidd was executed, Thomas Paine was around sixty-eight years old, but he had a good thirteen years left in him. In 1706, during Queen Anne's War, he went to sea again, in joint command of an expedition consisting of two ships and 120 men, dispatched to hunt down a French privateer from that old pirate haunt Petit Goâve. The venture was successful, and they brought the privateer back to Newport as a prize. At age seventy-three or thereabouts, it was the old buccaneer's final venture in armed conflict at sea. While one wonders if this too was one of his old comrades, he had probably outlived all of "the Men of Aves."

Paine lived out the rest of his years quietly in Jamestown. He died in the spring of 1715. He was in his eighties, a good long life for one who had lived so hard and fast, exposed himself to the dangers of sea and sword, fever and the rope, arrest and flying shot. His wife, Mercy, died three years later, and they are buried together on their property on Conanicut Island. Their house still stands to this day.

A pirate's retirement: Thomas Paine's home in Rhode Island

Thomas Paine, one of the last of the old-time buccaneers, was gone. His death came two years after the Peace of Utrecht, which ended the War of the Spanish Succession in Europe. That cessation of hostilities threw countless sailors and privateersmen out of work and sparked the last great wave of piracy in the Caribbean.

Paine did not live to see the high point of the eighteenth-century pirates, Blackbeard, Bart Roberts, Bellamy, and others. He was a man of an earlier era, and he was gone before this second great pirate awakening. Gone, but his fingerprints remained.

He and Mercy had no children. His namesake appears to be his nephew, Thomas Paine of Block Island. In 1718, Thomas became the second husband of Elizabeth (Williams) McCarty. Elizabeth was a sister of Palgrave Williams of Newport, Rhode Island. And, as readers of *Expedition Whydah* will recall, Palgrave Williams was the partner of Samuel Bellamy, pirate captain of the *Whydah Galley*.

And so it had come full circle. From the brandished forearm of Cape Cod to the deadly scorpion-tail reef of Las Aves, a chain stretched four decades between two sunken graveyards. I could picture the old buccaneer by the fireside during the long nights of winter, filling the ears of two ambitious young men with stories of shipwrecks and sunken treasure. I remembered what Jim Hawkins said of Billy Bones's tales in *Treasure Island:* "Dreadful stories they were; about hanging, and walking the plank, and storms at sea, and the Dry Tortugas, and wild deeds and places on the Spanish Main. By his own account he must have lived his life among some of the wickedest men that God ever allowed upon the sea . . ."

That might serve well as epitaph for Thomas Paine.[2]

39

The Legacy of Las Aves

B y the time we were scheduled to leave Las Aves, we were ready.
We weren't arguing—not much, anyway—but we were still
ready to leave.

On any expedition, when people are thrown together in close quar-
ters night and day, in harsh and demanding conditions, they start to
chafe against one another. This trip was no exception.

The accelerated schedule only raised the stress level. On a longer
trip, there is more of an opportunity to get away from one another—
if only for a few hours. If someone gets sick, they can take a break for
a day or two. There is more time to satisfy different personal agendas
and needs. But we did not have those luxuries.

As we moved into the last few days of the project, Margot and I
moved from our cabin into one with better ventilation, and my flu
symptoms disappeared.

The Venezuelan navy and coast guard were still dogging us. As the
weather window closed, our fears of returning to the reef increased,
adding to the general tension.

In the evenings we would sit around the *Antares,* discussing the
day's work, studying videotapes for targets we may have overlooked,
and swapping stories and tall tales. Charles is a natural raconteur and

he has picked up some prize stories in the course of his adventures in the jungle. Once he told us how he returned to camp with a band of Yanomami Indians after a day of cutting wood in the jungle. That night, while they were sitting around the fire, the sound of chopping hardwood still resonated eerily through the blackness. Charles, somewhat unnerved, asked his Yanomami companions, "Do you hear that chopping sound?" "Yes. That is the sound of us chopping wood tomorrow."

"Ah, that explains it," Charles said.

He told us another story of a woman from New York who was part of a group he was guiding through the jungle. He told her not to wander off, but she did not follow instructions, and soon got lost. For three days she wandered alone in the jungle. Charles and the others eventually found her hiding in a hollow log. She was unrecognizable from insect bites and nearly mad with fear.

The woman told them how a jaguar had stalked her. She claimed that at night the cat would lie down beside the log in which she was hiding and whisper to her in Spanish.

"New Yorkers," Charles said with more than a hint of repugnance. "She should have listened to me. Ah, but she learned a good lesson."

Charles's ability to spin a tale made it obvious why he was such a popular lecturer with organizations like the New York Botanical Society. But there was darkness in his stories, a menacing darkness that made me thankful that we were in my element, the sea, rather than deep in the Amazon jungle.

Near the end of the expedition, Charles's adult daughter from a previous marriage and some of her friends came out for a visit, and our crew and those aboard Charles's boat had a rare get-together. Charles's daughter owns a horse farm in the country. I found her delightful and charming.

Some of Charles's friends had come out as well. They were all from the upper crust of Venezuelan society. Like Charles, they were pleasant and affable.

Conversation flowed from rock stars to the relative toxicity of sea snakes to Nazis living in South America. But when the topic changed to politics, a shadow seemed to drift over our dinner guests, like the silhouette of a giant condor circling a flock of spring lambs. Indeed, our guests spoke of the prospect of Hugo Chávez as president as if he were planning to feature them as the main course at his inaugural dinner.

Charles bragged, "I can hold them off forever with a small, well-trained force and escape out my back door—the Amazon jungle is my backyard." He went on, "You Americans, you just don't know how to treat your blacks. Here, we know how to treat them. They know their place."

The room went silent. The ship's crew, all of whom were of African origins, looked at Charles—and us—with that faraway gaze that comes from generations of studied endurance in the face of calculated dehumanization.

I searched for words to defend my new friends but could only shake my head, too taken aback to open my mouth.

As our allotted time on the reefs dwindled, it was evident that we would not have time to locate and map every site as meticulously as we had been doing. The reef was four miles long and we could not cover it all. The best we could hope for was to make a quick visual survey, to see how closely the physical location of the remaining wrecks matched d'Estrées' map. The accuracy of ancient mapmakers has always been of interest to me. The exactitude of cartographers has often aided our efforts to find historic shipwrecks. For example, the precision of Captain Cyprian Southack's map of Cape Cod had played an important part in our discovery of the *Whydah*.

For our test of d'Estrées' map we decided to use the sled.

Basically a product of 1950s backyard technology, the sled, or hydroplane, looks like a large wooden cutting board with two grips that are essentially an upscale version of what a water skier holds on to when he is being towed behind a boat. And that is exactly the sled's function. The boat tows a sled through the water with a live passenger who is facedown, scanning the bottom.

Since the sled is hydrodynamic, it gives the person being towed a lot of maneuverability. The rider takes a big breath of air, then tilts the sled down to fly toward the bottom. He can tilt it up to fly back to the surface. With a turn of the wrist we could dive down or shoot back up.

Being almost laughably low-tech, the sled is often underutilized, but, even with limitations, it's a great way to check out large, shallow-water areas to get an overall picture of the seabed. It provided us with an extra day of searching that we otherwise would not have had—and let some of the team have a good time doing it.

Todd, Carl, Chris, and big Ron Hoogesteyn set out in the *Aquana* to search the north end of the reef with the sled.

There is a type of magnetometer called a proton precession magnetometer. The men referred to the sled as the "protein magnetometer," because, trolling along behind the boat, they felt like fish bait. In fact, Ron was wearing some jewelry around his neck that flashed in the sunlight coming down through the water and drew unwanted attention. When he discovered that a barracuda was trailing him as he trolled along on the sled, he quickly took it off. Ron had eaten plenty of barracuda in his life, and he did not want the roles reversed. For my part, I was impressed with the barracuda's appetite; Ron weighs about 260 pounds!

The crew spent an entire day searching the reef from the northernmost wreck we had located. Riding the sled is exhausting, but they liked that. They would take turns on it, switching at twenty-minute intervals. It became a competition, as most things did. The sled is a perfect arena for the endless competition between salvage divers, SEALs, and Special Forces divers to blow off steam and test one another's capacity for pain endurance.

At least they had fun, because the search was not successful. They combed the entire northern part of the reef and did not find a thing. It is quite likely that the wrecks were too embedded in the coral, too camouflaged, to spot by a quick glance from a towed diver. It would require a slower, more meticulous search to find the four wreck sites d'Estrées noted at the northern end of the reef.

For my part, I selfishly wanted to swim alone around the pirate shipwrecks. For me, the process of discovery is not exclusively scientific observation. The experience of simply being there *with* the wrecks was a window into the past in a way that is difficult to explain.

Each artifact from a wreck has its own story to tell of what life aboard that ship was like. And each artifact will tell its fascinating story as a part of the scientific process. But viewing an entire wreck in situ speaks volumes about life aboard ship—and the final moments of that ship.

Sometimes the whole *is* more than the sum of its parts. Hard science alone does not do justice to the cause of fully preserving the past. For that, you also need *heart,* a capacity for appreciating the drama and tragedy of a ship's dying moments. For an explorer, it is nearly as important to get at the meanings behind the data, as it is to gather the data itself.

The day before we were scheduled to wrap things up, we got some good news. Our permit from the navy had come through.

The BBC already had the film in the can. We had located and mapped nine of thirteen wrecks and proved the accuracy of d'Estrées' map. An adventure that began with Max Kennedy's tale of lost cannon had yielded a rich and rewarding prize: a lost fleet that virtually changed Caribbean history. That we had found two pirate vessels from the age of the buccaneers in the process was a bonus, especially for me.

Now it was time for us to go, and for the Venezuelan government to decide the fate of the sites we had found.

As we prepared to leave, Charles asked Mike if the *Antares* would take him and his gear back to Los Roques. This would save him the trouble of loading it all back on his own boat. It was a reasonable request: Los Roques was the *Antares's* destination as well, and we had plenty of room, but Mike refused outright.

I was surprised that Charles would ask Mike if he could come along, after all that had come between them. It was like Bill Clinton asking George W. Bush if he could hitch a ride to Arkansas on Air Force One's next trip to Texas. I tried to change Mike's mind, but he had had enough of Charles, and that was that.

We left Las Aves and cruised back to Los Roques. Yankey Willems had made the same trip three years after the loss of the French fleet. He had shown up at Las Aves, fished up two cannons off one of the wrecks, and brought them back to Los Roques, where he had careened. These had been Sam Bellamy's waters as well. So, once again, we were sailing in the wake of the buccaneers.

From Los Roques we took a plane back to Caracas. I spent a day there with Bart Jones, the AP reporter, before heading back to the United States. Timing is everything when you're south of the border. It was the day before the election, and the atmosphere was extremely tense. Bart took Margot and me to an overcrowded bar popular with foreign journalists and novelists such as Gabriel García Márquez. It was like stepping into the Hemingway novel *For Whom the Bell Tolls,* which is set on the eve of the Spanish Civil War.

Our explorations at Las Aves created quite a bit of attention in the press. Bart wrote several articles about the expedition, as did Charles Brewer. Charles's articles were in stark contrast to Bart's. Charles was out to lambaste Mike and me in the press, and, since Bart wouldn't join in, Charles declared war on Bart as well.

He also sent an e-mail to Max Kennedy describing me as "a good performer. A medicine man." Disappointed that I had not allowed the expedition to degenerate into a treasure hunt, he called me "an opportunistic pirate" and characterized my desire to see the integrity of the reef preserved as a ploy to cleanse myself "in front of the international opinion."

And so it goes. . . .

From Pinzón versus Columbus, through Speke versus Burton, to all the Mount Everest tiffs and *Darkness in Eldorado,* there are no rivalries in this world like the rivalries among explorers.

Mike Rossiter, in the meantime, tried to find out what had happened to delay the permits. He was concerned about what had taken place and wanted to be certain that there had been no official problems, nothing that would reflect badly on the BBC. He learned that virtually every claim Charles had made about the permits he had supposedly secured was untrue. It was like learning that a weather forecaster who knew a hurricane was on its way had nonetheless given "clear sailing" advice to a regatta.[1]

We learned that many people in the Venezuelan government had worked hard to overcome the problem. High-level government people had lobbied for us in Caracas. They had dealt mostly with Antonio Casado, who spoke Spanish and had led the way since he was in the city. When we returned to Caracas, we asked him what had happened, and how permission had been finally secured. It was a mystery to him as well: "I never really did understand. Suddenly I was told fine, the permit is granted."

Ironically, when our story hit the press, it was a perfect opportunity for the competing Venezuelan company to take the heat off their failed Nazi U-boat project by touting the Las Aves wrecks as a treasure fleet worth over $100 million. Some months later, after learning how well our expedition was conducted, one of their principals came to Cape Cod and asked me to lead another expedition to Venezuela for their company.

In any case, we were packed and headed for the airport and home. It was election day in November 1998, and soldiers with automatic weapons were everywhere, inspecting each car on its way to the airport. "A good day to get out of Venezuela," I thought.

And that ended my involvement with the wrecks on Las Aves. I like to think we managed to make some significant advances for maritime history. We identified nine wrecks—confirming the fundamental accuracy of d'Estrées' map. We had also made very accurate drawings of the wreck sites, recording a great deal of information that might otherwise be lost.

We also agreed, once the dust from the election settled, to lobby the Venezuelan government to protect Las Aves from vandals and to preserve it as a marine sanctuary where well-supervised sport divers could tour the site of one of the most interesting maritime disasters in the history of the Americas, on one of the most beautiful reefs in the world.

My entire background and experience is in locating wrecks, and in recovering, preserving, and displaying their remains. That's how I make my living—not by selling artifacts. But it is my opinion that the wrecks of the reef of Las Aves should be left untouched.

There has been no official decision concerning the disposition of the wrecks. But, with so many of its citizenry hungry and in rags, preserving ancient shipwrecks is not a priority of the Venezuelan government.

It has been reported that others have dived on the wrecks and have looted artifacts. Pedro Mezquita told the Sunday *Boston Herald,* "If the government does not take immediate action to protect the place there will be a new piracy." Pedro, more than most, knows how things go. "In Venezuela," he added, "national parks are not very well protected. Governments change. Policies change."

Governments themselves are sometimes the problem. There are suggestions that our permit problems were more than just bureaucratic red tape. There may have been other people with government connections with an eye on Las Aves, who were trying to block our progress.

It cannot be assumed that all of these were treasure hunters. It has been my unfortunate experience that the "ethic" of some archaeologists is geared far more toward the past than to the needs of the present or future.

The reefs at Las Aves are among the most beautiful ecosystems I have ever seen. D'Estrées' ships are no longer just wreckage lying atop the reef waiting to be excavated. After three hundred years, they *are* the reef, the living reef, inextricable parts of the whole. Even the

Margot Nicol-Hathaway and Barry Clifford exploring the reef

most careful and unobtrusive archaeological excavation would cause irreparable damage to a natural resource.

These great ships should be allowed to fade away, like the bones of the men who sailed them, until they too are no more than history, a part of the endlessly fascinating legacy of the Spanish Main.

May the coral be their tombstone.

Notes

Chapter 1

1. "A list of ffrench fleet which was under ye Comand of Count d'Estrée and designed for Curacoa," Colonial Office Papers (hereinafter cited as CO), British Public Record Office, London, 142 no. 98XV.
2. William Dampier, *A New Voyage Round the World* (London: Adam & Charles Black, 1937).
3. "Governor Stapleton to Lords of Trade and Plantations," Nevis, April 29, 1678, 10:690, Calendar of State Papers: Colonial Series; America and the West Indies, Great Britain Public Record Office, London. Her Majesty's Stationery Office 1860–1969 (hereinafter cited as CSPCS).
4. Ibid.
5. Quoted in David F. Marley, *Pirates and Privateers of the Americas* (Santa Barbara, Calif.: ABC-CLIO, Inc., 1994), p. 137.

Chapter 2

1. Alexandre Exquemelin, *The Buccaneers of America* (Glorieta, N. Mex.: The Rio Grande Press, 1992; reprint of 1684 edition), p. 103.
2. A fire ship is a small vessel built to be set on fire. Fire ships were sailed into enemy anchorages and ignited in the hope that they would set the anchored vessels on fire. They rarely did, though they often managed to create a panic that led to chaos and destruction in its own right.

CHAPTER 5

1. Dampier, *A New Voyage,* p. 43.
2. C. H. Haring, *The Buccaneers in the West Indies in the XVII Century* (London: Methuen, 1966) reprint of 1910 edition, pp. 220–221. Primary sources vary in the details of this event and not all of them mention this warning.
3. Freebooter is in part derived from the term "free booty," i.e., stolen goods, but the origins of the word go back further. Originally derived from the Dutch *vrijbuiter,* it has the same origin as *flibustier,* the French term that became filibuster, or pirate. All the terms mean, in essence, a robber, though they soon came to mean more specifically a pirate. Buccaneer, *flibustier,* filibuster, freebooter, and pirate were all used synonymously to mean the seaborne raiders of the Caribbean and Spanish Main.
4. Dampier, *A New Voyage,* p. 44.

CHAPTER 6

1. "Governor Sir Jonathan Atkins to Secretary Coventry," Barbados, August 1, 1678. CSPCS, Addenda Volume, 10: 1646.
2. Ibid.
3. Dampier, *A New Voyage,* p. 44. Thirty pounds was more than a year's income for most people in the seventeenth century.

CHAPTER 8

1. Patrick Tierney's *Darkness in Eldorado: How Scientists and Journalists Devastated the Amazon* (New York: W. W Norton and Company, 2000) covers in great detail all of the shocking events surrounding the exploitation of the Yanomami. Charles Brewer's participation is chronicled in depth.

CHAPTER 9

1. Maurice Besson, *The Scourge of the Indies: Buccaneers, Corsairs and Filibusters* (New York: Random House, 1929), p. 49.

CHAPTER 10

1. Exquemelin, *Buccaneers of America*, p. 89.

CHAPTER 13

1. Dampier, *A New Voyage,* p. 44. Interestingly, a pirate of a later generation, Calico Jack Rackam, would pull an almost identical ruse off the coast of Cuba. Trapped by a Spanish *guarda del costa,* Rackam and his men rowed out to the Spaniard's prize, an English sloop, took her, and sailed away in the night, leaving, according to Charles Johnson's *A General History of the Pirates* (Manuel Schonhorn, editor [Columbia, S.C.: University of South Carolina Press, 1972], p. 149), "but an old crazy hull in the room of her."
2. "Don Pedro de Ronquillos, Spanish ambassador to the King," Windsor, Sept. 6, 1680. CSPCS 10:1497.
3. Ibid., 1498.
4. Ibid.

CHAPTER 14

1. Haring, *The Buccaneers in the West Indies,* p. 241.
2. "Relation de la prise de la Gouaire," Archives Coloniales, F3 162, fol. 132. I am indebted to Raynald Laprise for this reference and translation.
3. Dampier, *A New Voyage,* p. 28.
4. Ibid., p. 30.
5. Ibid.
6. Ibid., p. 35.
7. "Sir Thomas Lynch to Secretary Sir Leoline Jenkins," Jamaica, Nov. 6, 1682. CSPCS 11:769.
8. The translation is "through fair means or foul."

CHAPTER 17

Much of this chapter is based on Amy Turner Bushnell, "Pirates March on St. Augustine," *El Eseribano,* April 1972, pp. 51–72.

1. Sir Thomas Lynch, quoted in Marley, *Pirates and Privateers,* p. 304.

2. "The King to the Governor and Magistrates of Massachusetts," Nevis [?], April 13, 1684. CSPCS 11:1634.

CHAPTER 18

1. "Earl of Craven to Lords of Trade and Plantations," May 27, 1684. CSPCS 11:1707.
2. Governor Edward Cranfield, quoted in Marley, *Pirates and Privateers,* p. 304.
3. "Relation of T. Thacker, Deputy Collector," Boston, August 16, 1684. CSPCS 11:1862ii.
4. Ibid.
5. Ibid.
6. "Lynch to Jenkins," CSPCS 11:769.
7. "The King to Sir Thomas Lynch," Windsor, April 13, 1684. CSPCS 11:1633.
8. Alexander Boyd Hawes, *Off Soundings: Aspects of the Maritime History of Rhode Island* (Chevy Chase, Md.: Posterity Press, 1999), p. 11. Hawes uncovered this information in the Public Record Office in London. He also uncovered a record in the Jamaican archives from 1689 that designates Lynch as a Gentleman of the Privy Council.
9. William Dyer, quoted in Howard M. Chapin, "Captain Paine of Cajacet," *Rhode Island Historical Society Collections,* January 1930, vol. 30, no. 1, p. 23.

CHAPTER 21

1. Quoted in Marley, *Pirates and Privateers,* p. 105.
2. "Sir Henry Morgan to the Lords of Trade and Plantations," St. Jago de la Vega, July 2, 1681. CSPCS 11:158.
3. Quoted in Marley, *Pirates and Privateers,* p. 105.
4. "Morgan to Lords of Trade and Plantations," CSPCS 11:158.
5. "Symon Musgrave to [Governor Sir Thomas Lynch]," Port Royal, Sept. 29, 1682. CSPCS 11:709.
6. Ibid.

CHAPTER 22

1. "Affidavits of Van Hoorn's Piracies. Depositions of James Nicholas, gunner; John Otto, coxswain; Peter Cornelius, sailmaker; George

Martyn, sailor, late of the ship *Mary and Martha,* alias *St. Nicholas,* 400 tons, 40 guns," March 3, 1683. CSPCS 11:963i.

2. Ibid.
3. "Sir Thomas Lynch to William Blathwayt," Jamaica, Feb. 22, 1683. CSPCS 11:963.
4. Ibid.
5. Ibid. In the last sentence, Lynch is referring to the fiction of Van Hoorn's being sent after pirates on Ile à Vache, an island off the southern coast of Haiti. Had he been going to Ile à Vache, Van Hoorn would not have carried six months' worth of provisions. Nor would he have sailed all the way downwind to Jamaica ("come to leeward"), which would have forced him to make the tedious and difficult sail back against the trade winds to fetch Haiti ("when he knows they are to windward").
6. Ibid.

CHAPTER 25

1. "Lynch to Blathwayt," CSPCS 11:963.
2. "The Examination and Confession of Robt. Dangerfield aged thirty-two years or thereabouts taken this 27 Sept 1684," CO 1/057 146 ff. 375–376 (565–568).
3. "Sir Thomas Lynch to Secretary Sir Leoline Jenkins," Jamaica, July 26, 1683. CSPCS 11:1163.
4. "Sir Thomas Lynch to the Lord President of the Council," Jamaica, May 6, 1683. CSPCS 11:1065.
5. De Grammont was most likely in his forties at this time, which was getting up in age for a buccaneer.
6. Little is known about Foccard's activities prior to this, though he would go on to participate in some of the major pirate raids of the 1680s.

CHAPTER 26

1. "Captain Van Hoorn's Taking of Vera Cruz," in *The Voyages and Adventures of Capt. Bart Sharp and others in the South Seas: being a Journal of the same* [Anonymous]. Printed by Philip Ayres, 1684.
2. "Sir Thomas Lynch to the Lord President of the Council," Jamaica, May 6, 1683. CSPCS 11:1065.

3. The same mistake allowed John Hawkins to sail unopposed into Vera Cruz over one hundred years before.

4. Cochenelle, or Cochineal, a scarlet dye, consists of the dried bodies of the insect *coccus cacti,* which is found on several species of Mexican cactus. Like indigo, it was extremely valuable. One can scarcely imagine the labor involved in collecting enough dried insect bodies to fill a two-hundred-pound bag.

5. "Captain Van Hoorn's Taking of Vera Cruz."

6. "Sir Thomas Lynch to the Lords of Trade and Plantations," Jamaica, Feb. 28, 1684. CSPCS 11:1563.

CHAPTER 29

1. "Governor Sir Thomas Lynch to the Governor of Havana," Jamaica, Aug. 18, 1683. CSPCS 11:1198.

2. Ibid.

3. Ibid.

4. "The pirate Laurens to Governor Sir Thomas Lynch," Petit Gouaisne [*sic*], Aug. 24/Sept. 3, 1683. CSPCS 11:1210.

5. Once again, Lynch's instincts seem almost uncanny. Nearly a year later there would be much urging among the buccaneers for another attempt on Vera Cruz.

CHAPTER 30

1. "Lynch to the Lords of Trade and Plantations," CSPCS 11:1563. Lynch, like many people of his time, used the terms "privateer" and "pirate" almost interchangeably, though strictly speaking a pirate operated with no official commission. For practical purposes during the seventeenth century, it seems a privateer was one who attacked other countries' ships, with or without a commission, but left yours alone.

2. Ibid.

3. "Laurens the pirate to Sir Thomas Lynch," St. Philip's Bay, April 26/May 6, 1684. CSPCS 11:1649.

4. "Sir Thomas Lynch to the pirate, Laurens," Jamaica, August 15, 1684. CSPCS 11:1839.

5. "Sir Thomas Lynch's Overtures to the pirate, Laurens," Jamaica, May 31, 1684. CSPCS 11:1718.

6. Governor Edward Cranfield, quoted in Marley, *Pirates and Privateers,* p. 13.

7. All quotes from Marcus Rediker, *Between the Devil and the Deep Blue Sea* (Cambridge, U.K.: Cambridge University Press, 1998), pp. 64–66.

8. Bradstreet, who was too tolerant of pirates for the Crown's taste, was eventually replaced by Edmund Andros, a Crown appointee. Andros, however, was too strict about enforcing Crown policy for the colonists' taste. In 1689 he was ejected by the colonists in what has been called "America's First Revolution."

9. The story of de Graff's marriage is from C. H. Haring, *Buccaneers in the West Indies,* p. 246. Haring quotes E. Ducere's "Les Corsaires sous l'ancien regime," Bayonne, 1895.

10. Ravenau de Lussan, *Journal of a Voyage into the South Seas* (Cleveland: Arthur H. Clark, 1930), p. 7.

11. Ibid.

12. "Sir Thomas Lynch to the Lord President of the Council," Jamaica, June 20, 1684. CSPCS 11:1759.

CHAPTER 31

1. Ironically, Joseph Bannister's defection to the French appears to have hastened Lynch's death. Bannister had been tried for piracy, but the Jamaican jury acquitted him of the charge. Lynch, who had not been well for some time, was apparently so infuriated by that decision that his anger pushed him over the brink. He died a week after the verdict was handed down.

2. "Lt. Governor Molesworth to William Blathwayt," Jamaica, May 15, 1685. CSPCS 12:193.

3. Ibid.

4. Quoted in Robert S. Weddle, *Wilderness Manhunt: The Spanish Seach for La Salle* (Austin, Tex.: University of Texas Press, 1973), p. 40.

5. Quoted in Marley, *Pirates and Privateers,* p. 167.

6. Haring, *Buccaneers of the West Indies,* p. 246.

CHAPTER 32

1. Tierney, *Darkness in Eldorado,* p. 157.

2. Ibid., pp. 155–57, 193.

CHAPTER 34

1. Andrés de Pez was one of the few officers involved who did not receive censure from the court-martial. His small vessel could not have been expected to take part in a fight between the heavy hitters.
2. "De Cussy to Molesworth," quoted in Marley, *Pirates and Privateers,* p. 113.
3. Ibid.
4. "Lt. Governor Molesworth to William Blathwayt," Jamaica, Oct. 4, 1687. CSPCS 12:1450.
5. Ibid.

CHAPTER 35

1. De Cussy, quoted in Marley, *Pirates and Privateers,* p. 114.
2. "Governor the Duke of Albemarle to Lords of Trade and Plantations," Jamaica, Aug. 8, 1688. CSPCS 12:1858.
3. "Sir Francis Watson to Lords of Trade and Plantations," Jamaica, Apr. 22, 1689. CSPCS 13:85.
4. "Sir Francis Watson to Lords of Trade and Plantations," Jamaica, Oct. 27, 1689. CSPCS 13:515.
5. De Cussy, quoted in Marley, *Pirates and Privateers,* p. 96.
6. "Minutes of the Council of Jamaica," Dec. 3, 1689. CSPCS 13:628.
7. "Minutes of the Council of Jamaica," Dec. 9, 1689. CSPCS 13:628.
8. "Minutes of the Council of Jamaica," Dec. 12, 1689. CSPCS 13:635.
9. "Earl of Inchiquin to Lords of Trade and Plantations," Jamaica, July 6, 1690. CSPCS 13:980.
10. De Cussy, quoted in Marley, *Pirates and Privateers,* p. 115.

CHAPTER 38

1. The quotes in this chapter are from Samuel Niles's account of the battle as given in Chapin, "Captain Paine of Cajacet."
2. In 1891, workmen reportedly discovered "a quantity of elephant tusks, silver coins and gold doubloons" while digging under Paine's house, then owned by the Vose family. Robert S. Cahill, *New England's Pirates and Lost Treasures* (Peabody, Mass.: Chandler Smith, 1987), p. 22.

CHAPTER 39

1. I later learned that, in one instance, Charles had begun gold-mining operations six years before his permits went into effect, and that in 1984 he had been arrested by the Venezuelan national guard for gold mining in a prohibited area and exporting Venezuelan fauna while using unsalaried Indians as workers. See Tierney, *Darkness in Eldorado,* pp. 153–56.

Acknowledgments

The success of any underwater archaeological survey expedition depends on the help of many individuals and organizations. Aside from the members of the actual project team such as Todd Murphy, Chris Macort, Cathrine Harker, Eric Scharmer, Carl Tiska, and Margot Nicol-Hathaway, I also want to thank Antonio Casado, Pedro Mezquita, Dr. John de Bry, and especially Max Kennedy for sharing with me his boyhood dream of finding the Lost Fleet.

Credit for bringing the Lost Fleet from the reefs of Las Aves to the world's living rooms is due to the BBC and the Discovery Channel. Special thanks are owed to Mike Quattrone, Mike Rossiter, Rebecca Lavender, and the production teams.

In many ways, a book is like an expedition—it is never the work of one person. I would like to thank my agent, Nat Sobel; Jim Nelson for his work; Diane Reverend and Dan Conaway for their encouragement and support; and Matthew Guma and Nikola Scott for their cheerful and unflagging efforts in the book's production.

Most especially, my gratitude to Ken Kinkor, project historian, comrade, and friend, without whom this book would not have been possible.

Credits

page   2: Barry Clifford collection with thanks to John de Bry

3: N. Ozarne

4: Artist unknown; from David F. Marley, *Pirates: Adventurers of the High Seas,* London: Arms and Armour Press, 1995, p. 69

6: Pablo Tillac

8: From Maurice Besson, *The Scourge of the Indies: Buccaneers, Corsairs and Filibusters*, New York: Random House, 1929

9: From Alexandre Exquemelin, *The Buccaneers of America*, Glorietta, N. Mex.: Rio Grande Press, 1992; reprint of 1684 edition

10: From Besson, *The Scourge of the Indies*

11: Howard Pyle

13: From Exquemelin, *Buccaneers of America*

14: From Exquemelin, *Buccaneers of America*

16: Henry B. Culver

17: Barry Clifford collection with thanks to John de Bry

18: F. Perrot

20: Paul Ryan

24: Barry Clifford

26: Barry Clifford

27: Barry Clifford

28: Barry Clifford

29: Paul Ryan

32: J. C. Shetky

34: Barry Clifford

38: Howard Pyle

39: Barry Clifford

43: Barry Clifford

48: Barry Clifford

49: Barry Clifford

50: Barry Clifford

page 52: Margot Nicol-Hathaway
54: D'Estrées' map
56: From Besson, *The Scourge of the Indies*
65: Pouget
75: Margot Nicol-Hathaway
81: From Captain Charles Johnson, *A General History of the Pirates,* London, 1734
84: Henry B. Culver
89: Artist unknown; from C. H. Haring, *Buccaneers in the West Indies in the XVII Century,* London, 1966, reprint of 1910 edition
91: Henry B. Culver
102: Paul Ryan
107: Paul Ryan
113: Howard Pyle
120: Margot Nicol-Hathaway
121: Margot Nicol-Hathaway
124: Margot Nicol-Hathaway
132: Randon
135: Artist unknown; from Johnson, *A General History of the Pirates*
136: From Exquemelin, *Buccaneers of America*
139: Henry B. Culver
142: William van de Velde the Younger
143: Pablo Tillac
149: Chris Macort
153: Margot Nicol-Hathaway
159: William van de Velde the Elder
166: Howard Pyle
167: Howard Pyle
174: Paul Ryan
181: Henry B. Culver
183: Artist unknown; from Johnson, *A General History of the Pirates*
184: Artist unknown; Kenneth J. Kinkor collection
192: Artist unknown; from *The Great Age of Sail,* Switzerland: Edita Lausanne, 1967, p. 86
194: From Besson, *The Scourge of the Indies*
197: Henry B. Culver
201: Pablo Tillac
202: From Besson, *The Scourge of the Indies*
203: From Exquemelin, *Buccaneers of America*

page 211: Paul Ryan

213: Margot Nicol-Hathaway

214: Barry Clifford

216: Margot Nicol-Hathaway

217: Paul Ryan

222: Margot Nicol-Hathaway

227: I-V. Beecq; from Besson, *The Scourge of the Indies*

230: W. Woollett; from Frank C. Bowen, *Wooden Walls in Action*, London: Hutton & Co., 1951, p. 26

231: Artist unknown; from Howard Engel, *Lord High Executioner: An Unashamed Look at Hangmen, Headsmen, and their Kind*, Buffalo, N.Y.: Firefly Books, 1990

236: From Besson, *The Scourge of the Indies: Buccaneers*

239: Barry Clifford collection with thanks to John de Bry

242: Howard Pyle

245: Paul Ryan

246: Paul Ryan

249: Barry Clifford

251: Margot Nicol-Hathaway

254: Paul Ryan

263: From "Captain Paine of Cajacet," Howard M. Chapin, *Rhode Island Historical Society Collections,* January 1930, vol. 30, no.1

272: Paul Ryan